ROUTLEDGE LIBRAR
HUMAN GEOGi

T0227514

Volume 9

THE GEOGRAPHY OF HEALTH SERVICES IN BRITAIN

THE GEOGRAPHY OF HEALTH SERVICES IN BRITAIN

ROBIN HAYNES

Routledge
Taylor & Francis Group

LONDON AND NEW YORK

First published in 1987 by Croom Helm Ltd

This edition first published in 2016
by Routledge
2 Park Square, Milton Park, Abingdon, Oxon OX14 4RN

and by Routledge
711 Third Avenue, New York, NY 10017

Routledge is an imprint of the Taylor & Francis Group, an informa business

British Library Cataloguing in Publication Data
A catalogue record for this book is available from the British Library

ISBN: 978-1-138-95340-6 (Set)
ISBN: 978-1-315-65887-2 (Set) (ebk)
ISBN: 978-1-138-95477-9 (Volume 9) (hbk)
ISBN: 978-1-315-66677-8 (Volume 9) (ebk)

Publisher's Note
The publisher has gone to great lengths to ensure the quality of this reprint but points out that some imperfections in the original copies may be apparent.

Disclaimer
The publisher has made every effort to trace copyright holders and would welcome correspondence from those they have been unable to trace.

The Geography of Health Services in Britain

Robin Haynes

CROOM HELM
London • Sydney • Wolfeboro, New Hampshire

Croom Helm Ltd, Provident House, Burrell Row,
Beckenham, Kent BR3 1AT
Croom Helm Australia, 44–50 Waterloo Road,
North Ryde, 2113, New South Wales, Australia.

British Library Cataloguing in Publication Data

Haynes, Robin M.
  The geography of health services in
  Britain.
  1. National Health Service (Great Britain)
  2. Regional medical programs—Great
  Britain
  I. Title
  362.1'0941   RA395.G6
  ISBN 0–7099–3766–0

Croom Helm, 27 South Main Street,
Wolfeboro, New Hampshire 03894–2069, USA

Library of Congress Cataloging-in-Publication Data

Haynes, Robin M.
  The geography of health services in Britain.

  Bibliography: p.
  Includes index.
  1. National Health Service (Great Britain) —
Administration. 2. Medically underserved areas —
Great Britain. 3. Public health personnel —
Great Britain — Supply and demand. 4. Hospitals —
Great Britain — Location. 5. Medical care —
Great Britain. I. Title.
  RA412.5.G7H34 1987   362.1'0941   86-24250
  ISBN 0-7099-3766-0

Typeset by Florencetype Ltd, Bristol
Printed and bound in Great Britain
by Billing & Sons Limited, Worcester.

Contents

Figures

Tables

Preface

This book is about geographical variations in the organisation, provision and use of health services in Britain. Its main theme is that neither the quantity nor the quality of health care provided by the National Health Service is uniform from place to place. Local anomalies persist partly because of the way health services are administered. Chapter 1 traces the double reorganisation of the National Health Service in recent years and describes the present geographical pattern of administration. Chapter 2 explores the relationships between the need for health care and the supply of health services, to identify those parts of Britain which are relatively under-provided. The current policy for redistributing health service resources geographically is then evaluated. Chapters 3 and 4 examine the distribution of doctors, dentists, community nurses and hospitals over the country, looking for significant variations in coverage. The scope for future change is discussed. Chapter 5 is concerned with how people get to health services, whether by ambulance or other transport, and how the present system may create difficulties of access in particular circumstances. Difficulties in obtaining health services are most likely to be encountered in two contrasting types of area, remote rural areas and inner-city districts, for very different reasons. Chapters 6 and 7 describe the characteristics of people most at risk in such areas and review the policies available to ameliorate the problems. Chapter 8 describes the distribution of both private health facilities and social welfare services and discusses their effect on the National Health Service. Questions of priority and proposals for change occupy the final section, concerned with how the goal of social and geographical equity might be approached.

Several people have helped me to prepare this work. I wish particularly to thank Pauline Blanch and Sue Winston for typing the manuscript, Brett Davies for drawing the figures and Graham Bentham and Michael Bradford for valuable comments.

I am grateful to the following for giving me permission to use copyright material: The Open University for Figures 1.1,

1.2 and 1.6; the Controller of Her Majesty's Stationery Office for Figures 2.1, 2.2, 2.4, and 8.2; Dr D.R. Phillips for Figures 3.1, 3.2 and 4.2; Pion Ltd for Figure 3.4; Dr J.D. McGarrick and the British Dental Journal for Figure 3.5; Dr L.P. Grime and John Wright Journals for Figure 4.3; Policy Journals for Figure 5.2; Professor B. Jarman and the British Medical Journal for Figure 7.2; and Dr J. Mohan and Edward Arnold (Publishers) Ltd for Figure 8.1.

1

The Geographical Basis of Organisation

Does it matter where you live? For the majority of health service users in Britain, the question may come as a surprise. The National Health Service was established to provide a uniform standard of service for all, and its founders intended the nationalised system to ensure that 'an equally good service is available everywhere' (Bevan, 1945). To a great extent, the plan succeeded. Removing the financial barriers to access was a potent equalising force, yet, forty years later, some inequalities remain. Among these are the inequalities of location. The quality and quantity of health services available are not uniform geographically.

While most people in Britain today are able to consult their doctor in a properly appointed surgery or in a purpose-built health centre with nurses, dentists and pharmacists on hand, a small minority receive medical attention in shabby and unsuitable premises lacking even a water supply. An emergency call about a child's illness in the night is less likely to be answered by the family's own doctor in some areas than others. Preventive services which protect life, such as cervical screening, and those like chiropody, which preserve the quality of life, are not uniformly available. Waiting lists for repair operations may be discouragingly long in one place, yet in the next a person with the same condition may receive prompt attention. Patients admitted to hospital might find themselves in a new, technologically advanced complex, in a small local unit with limited facilities or in an obsolete Victorian edifice. Accident victims might benefit from rapid treatment in a 24-hour centre, or they might face a long wait and then an extended journey by ambulance, depending on where they live. In places with no local public transport, getting to the doctor, the dentist or the hospital is a real problem for some people. The ease of obtaining almost any sort of health service varies with geographical location.

Part of the reason for the differences in health services from place to place lies in the way the services are organised and administered. Some services are planned by health authorities, each of which has considerable autonomy within a relatively small district. Health authorities exist to provide services to suit local health needs, but there is no agreed method of determining needs. Authorities are constrained by the inheritance of existing buildings and local professional interests, so differences in the mix of services supplied persist across their boundaries. Other health services are outside the control of health authorities. Some are hardly planned at all, except in a very limited sense, and these have developed according to their own internal logic. There are serious problems of co-ordination between different branches of the health services. Weaknesses in one branch are not necessarily balanced by strengths in another, because there is no single body responsible for the overall effectiveness of local health care provision.

This first chapter is devoted to the administrative structure of the National Health Service, the framework within which most formal health services in Britain are planned and supplied. While private health services have grown rapidly in recent years, they are still small by comparison. The National Health Service has been reorganised twice recently, so a historical perspective is needed to understand the current system.

THE DEVELOPMENT OF NHS ORGANISATION

The early years

The present organisation of health services in Britain took its shape when the National Health Service (NHS) came into being in 1948 (Levitt and Wall, 1984). Figure 1.1 shows the arrangements for England. The NHS had a tripartite structure,

Figure 1.1: Organisation of health services in England in 1948

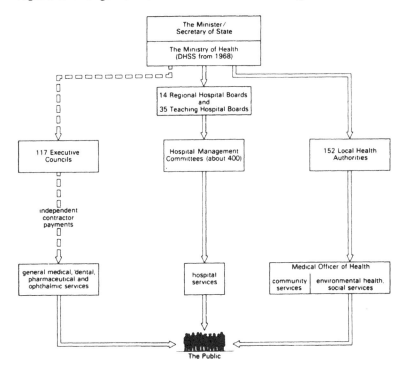

Source: Open University, 1985, p. 5, © The Open University Press.

with the three divisions administered by executive councils, hospital boards and local authorities.

One of the main difficulties facing the architects of the NHS was the reluctance of the medical profession to lose its traditional independence. The unenthusiastic co-operation of the British Medical Association was won only after general practitioners were guaranteed professional freedom. General medical practitioners, dental practitioners, pharmacists and opticians remained as independent contractors, working for themselves but paid fees by the NHS for their services. Payments were administered by executive councils made up of medical as well as lay members. Executive council areas roughly corresponded in size with local authority areas.

Hospital services, in contrast, were brought directly under the control of the Ministry of Health. Former municipal

3

hospitals and voluntary hospitals in the same vicinity were grouped together and each group was administered by a hospital management committee. The committees were responsible for hospitals and not for geographical areas, so it was not unusual for two or more hospital groups to provide complementary or overlapping services for the same population. Overall co-ordination and planning at the regional level were done by regional hospital boards, of which there were 15 in England and Wales, and 5 in Scotland. In England and Wales (but not Scotland) teaching hospitals were administered by boards of governors responsible to the Minister of Health and they were independent of the regional system. While these arrangements brought most hospitals into a nationalised organisation, some remained outside the NHS. These included certain religious institutions and commercial nursing homes.

The third set of services, the community and environmental health services, remained the responsibility of local government authorities (the councils of counties and county boroughs). Domiciliary midwifery, home nursing, health visiting, the welfare of mothers and young children, vaccination and immunisation, ambulance services, care of the mentally ill and mentally handicapped, the school health service, public health services and preventive health services were all in this category.

The advantage of leaving the community and environmental health services under the control of local authorities was that they could be co-ordinated with other local authority services such as personal social services, education and housing, whose objectives and interests were similar. The hospitals and family practitioner services, on the other hand, were deliberately kept out of local government control to make geographical uniformity more easily attainable. It was argued that if all health services depended on local rates there would tend to be a better service in the richer areas and a worse service in the poorer (Buxton and Klein, 1978). While there was no politically acceptable way to integrate general practitioner services fully into the nationalised system, their autonomy remains a source of difficulty for integrated planning. This is now a matter for concern, especially in the inner cities, as Chapter 7 will show.

Figure 1.2: Organisation of health services in England in 1974

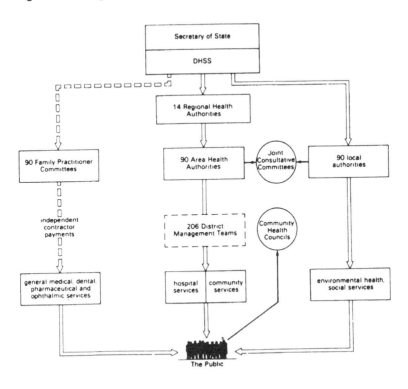

Source: Open University, 1985, p. 7, © The Open University Press.

The 1974 reorganisation

The problems of co-ordination between the three branches of the health services were not merely an administrative inconvenience. The lack of effective liaison between general practitioners, hospitals and community health services caused unnecessary delays and suffering for patients. There were increasing demands for unification (Medical Services Review Committee, 1962). At the same time, the fragmented system of local government came under review and the advantages of establishing a smaller number of local authorities emerged. It was decided to reorganise health services in parallel. The National Health Service Reorganisation Acts were passed by

5

Figure 1.3: Regional health authorities

a Conservative administration in 1972 and 1973 and implemented by a Labour government in 1974. In essence, the 1974
reorganisation brought the hospital and community health
services into an administrative structure consisting of three
hierarchical levels: regions, areas and districts (Figure 1.2).
These changes were timed to coincide with local government
reorganisation, except in Scotland, where the corresponding
local government changes were made one year later.

From 1974 the hospital regions were largely preserved as
administrative units, but each was headed by a regional health
authority with much broader responsibilities than the former
hospital boards for the planning and co-ordination of health

Figure 1.4: Scottish health boards

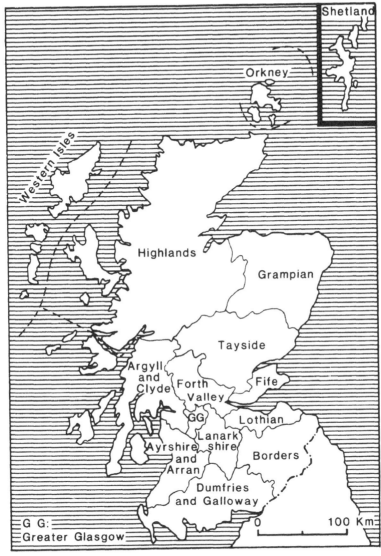

services in general. Regional health authorities (RHAs) are made up of members (the policy makers) and officers (employed to advise and carry out the policy), as in local government. Most of the members are lay people, not professionally connected to the health service. The officers consist of medical, nursing and administrative staff, headed more recently by a general

7

Figure 1.5: Health and social services boards in Northern Ireland

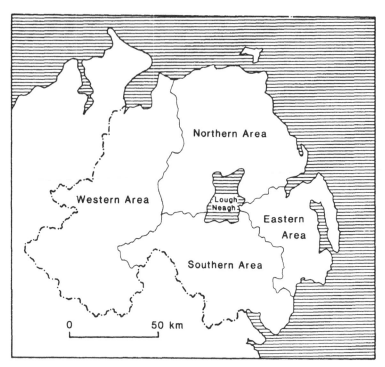

manager (DHSS, 1984d). Each region covers a major portion of the country, with populations ranging between 2 and over 5 million (Figure 1.3).

All regions were divided into areas, whose geographical boundaries corresponded with the new counties and the new local government districts in metropolitan counties. In London the health areas coincided with borough boundaries (themselves redefined in 1963). There had been some prior debate about whether health services should be administered within geographical units reflecting the local distribution of ill health, but the advantages of having one health service administrative body to match each local government authority prevailed. The health areas were administered by area health authorities. Collaboration with the matching local government authority was to be achieved through joint consultative committees. In addition to managing all NHS hospitals, including teaching hospitals, health authorities took over community health services (home nursing, the school service, ambulances and so on) for the first time. Local government retained its control over environmental, or 'public', health.

Scotland and Northern Ireland had no regional tier of management. The authorities established at area level are known as health boards. Scotland has 15 health boards, whose territories are shown in Figure 1.4. They administer all three branches of the NHS, including family practitioner services. Northern Ireland's four health and social service boards (Figure 1.5) have the unique responsibility of administering both health services and personal social services in the province.

Areas and health boards were themselves divided into districts, most of which approximated to the planned catchment of a district general hospital. It was at the district level that the day-to-day management of most health services took place, under the district management team consisting of a community physician, a nursing officer, an administrator and a finance officer. A further innovation was that in each district a community health council (CHC) was set up to represent the views of the public. In Scotland these bodies are called local health councils and district committees are the equivalent in Northern Ireland.

Family practitioner services remained separate under the 1974 reorganisation in England and Wales, but the executive councils were replaced by new family practitioner committees, each of which matched, and received its finance from, an area health authority.

Reorganisation introduced profound changes in the way health services were managed and also in their geographical integration. In Scotland and Northern Ireland, health services were brought under one administration. In England and Wales, reorganisation opened up the possibility of drawing the tripartite division closer together by allowing collaboration between matching family practitioner committees, health authorities and local government authorities at area level. It established a framework within which an overall health care strategy for a geographical area might be devised and implemented. These advantages were diluted when, in 1982, area health authorities were abolished.

Figure 1.6: Organisation of health services in England in 1985

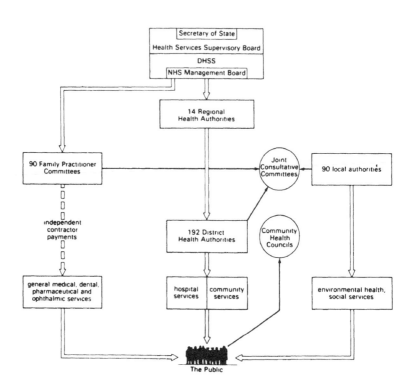

Source: Open University, 1985, p. 9, © The Open University Press.

More reorganisation

The reorganised hierarchy of the NHS had not been in existence for long before it came under criticism on the grounds that it was unnecessarily complicated, slow to take decisions, employed too many administrators and wasted money. To investigate these allegations and the financial difficulties facing the NHS the government appointed a Royal Commission on the National Health Service in 1976. Reporting in 1979, the commission took the view that the organisation did contain one tier too many and that the functions of areas and districts could be met more effectively by a single level of administration. The government translated this view into a set of proposals in a consultation document, *Patients First* (DHSS, 1979a) and then into policy (DHSS, 1980d).

The aim outlined in the policy was to leave the regional health authorities undisturbed and to establish under them one level of administration. In England, the area level was to be removed and replaced by district health authorities. Regional health authorities were given the job of proposing the division of their own regions into new districts, which they duly did, mostly basing the proposed district authorities on existing health districts or combinations of districts. After a short consultation period, the new district health authorities were announced in 1981 and came into being on 1 April 1982. Figure 1.6 shows the resulting organisational structure.

There are 191 health districts in England (Figure 1.7). Districts are intended as far as possible to be 'natural communities', focused on centres of population and linked by transport routes. The model district is an urban centre and its hinterland. Districts generally have populations of 100,000 to 400,000, although some urban districts exceed 500,000. The normal population range is equivalent to the catchment of one or two modern district general hospitals. Health district boundaries do not necessarily follow any other administrative divisions. District health authorities are responsible for the planning, provision and development of services within their boundaries. Like the regional authorities, they consist of appointed members to provide overall direction and employed officers for the daily management of services.

Within a health district, the lowest level of administration is the unit. A unit is a collection of services managed as one entity. A large hospital or several small hospitals, for example, could be a unit, as might be the community health services of the district, maternity services, or services for the mentally ill. Units are managed by a general manager working with a unit administrator, a director· of nursing services and senior member of the medical staff.

In Wales, the second round of reorganisation had a different effect. To keep the lowest level of administration similar in terms of resources used and population served to the new district health authorities in England, the generally smaller Welsh districts were removed and the Welsh area authorities retained. In the event, the eight Welsh area authorities became nine, and were renamed as district health authorities.

Figure 1.7: District health authorities

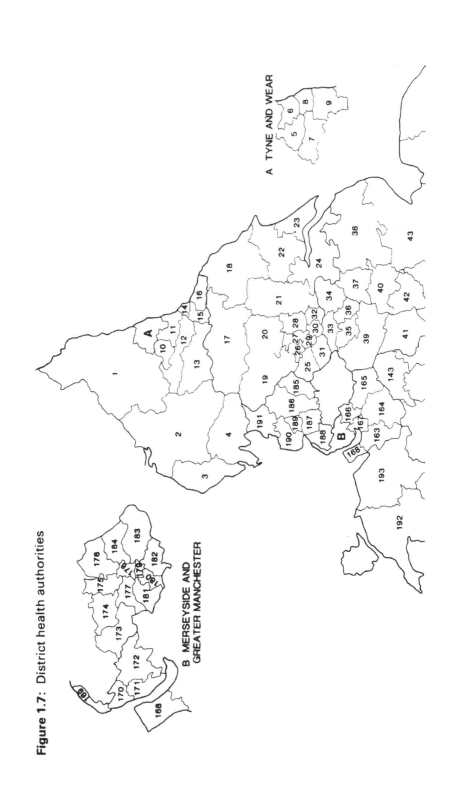

A TYNE AND WEAR

B MERSEYSIDE AND
GREATER MANCHESTER

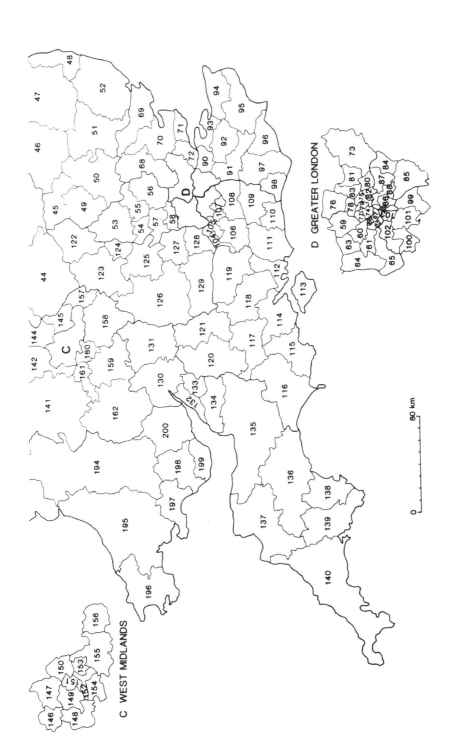

C WEST MIDLANDS

D GREATER LONDON

80 km

Key to Figure 1.7

Northern Region
1. Northumberland
2. East Cumbria
3. West Cumbria
4. South Cumbria
5. Newcastle
6. North Tyneside
7. Gateshead
8. South Tyneside
9. Sunderland
10. North West Durham
11. Durham
12. South West Durham
13. Darlington
14. Hartlepool
15. North Tees
16. South Tees

Yorkshire Region
17. Northallerton
18. Scarborough
19. Airedale
20. Harrogate
21. York
22. East Yorkshire
23. Hull
24. Scunthorpe
25. Calderdale
26. Bradford
27. Leeds Western
28. Leeds Eastern
29. Dewsbury
30. Wakefield
31. Huddersfield
32. Pontefract

Trent Region
33. Barnsley
34. Doncaster
35. Sheffield
36. Rotherham
37. Bassetlaw
38. North Lincolnshire
39. North Derbyshire
40. Central Nottinghamshire
41. Southern Derbyshire
42. Nottingham
43. South Lincolnshire
44. Leicestershire

East Anglian Region
45. Peterborough
46. West Norfolk and Wisbech
47. Norwich
48. Great Yarmouth and Waveney
49. Huntingdon
50. Cambridge
51. West Suffolk
52. East Suffolk

North West Thames Region
53. North Bedfordshire
54. South Bedfordshire
55. North Hertfordshire
56. East Hertfordshire
57. North West Hertfordshire
58. South West Hertfordshire
59. Barnet
60. Brent
61. Ealing
62. Hammersmith and Fulham
63. Harrow
64. Hillingdon
65. Hounslow and Spelthorne
66. Paddington and N. Kensington
67. Victoria

North East Thames Region
68. West Essex
69. North East Essex
70. Mid Essex
71. Southend
72. Basildon and Thurrock
73. Barking, Havering and Brentwood
74. Bloomsbury
75. City and Hackney
76. Enfield
77. Hampstead
78. Haringey
79. Islington
80. Newham
81. Redbridge
82. Tower Hamlets
83. Waltham Forest

South East Thames Region
84. Bexley
85. Bromley
86. Camberwell
87. Greenwich
88. Lewisham and N. Southwark
89. West Lambeth
90. Dartford and Gravesend
91. Tunbridge Wells
92. Maidstone
93. Medway
94. Canterbury and Thanet
95. South East Kent
96. Hastings
97. Eastbourne
98. Brighton

South West Thames Region
99. Croydon
100. Kingston and Esher
101. Merton and Sutton
102. Richmond, Twickenham and Roehampton
103. Wandsworth

104. West Surrey and N.E. Hampshire
105. North West Surrey
106. South West Surrey
107. Mid Surrey
108. East Surrey
109. Mid Downs
110. Worthing
111. Chichester

Wessex Region
112. Portsmouth and S.E. Hampshire
113. Isle of Wight
114. Southampton and S.W. Hampshire
115. East Dorset
116. West Dorset
117. Salisbury
118. Winchester
119. Basingstoke and N. Hampshire
120. Bath
121. Swindon

Oxford Region
122. Kettering
123. Northampton
124. Milton Keynes
125. Aylesbury Vale
126. Oxfordshire
127. Wycombe
128. East Berkshire
129. West Berkshire

South Western Region
130. Gloucester
131. Cheltenham and District
132. Southmead
133. Frenchay
134. Bristol and Weston
135. Somerset
136. Exeter
137. North Devon
138. Torbay
139. Plymouth
140. Cornwall and Isles of Scilly

West Midlands Region
141. Shropshire
142. Mid Staffordshire
143. North Staffordshire
144. South East Staffordshire
145. North Warwickshire
146. Wolverhampton
147. Walsall
148. Dudley
149. Sandwell
150. North Birmingham

151. West Birmingham
152. Central Birmingham
153. East Birmingham
154. South Birmingham
155. Solihull
156. Coventry
157. Rugby
158. South Warwickshire
159. Worcester and District
160. Bromsgrove and Redditch
161. Kidderminster and District
162. Herefordshire

Mersey Region
163. Chester
164. Crewe
165. Macclesfield
166. Warrington
167. Halton
168. Wirral
169. North Sefton
170. South Sefton
171. Liverpool
172. St. Helens and Knowsley

North Western Region
173. Wigan
174. Bolton
175. Bury
176. Rochdale
177. Salford
178. North Manchester
179. Central Manchester
180. South Manchester
181. Trafford
182. Stockport
183. Tameside and Glossop
184. Oldham
185. Burnley, Pendle and Rossendale
186. Blackburn, Hyndburn and Ribble Valley
187. Chorley and South Ribble
188. West Lancashire
189. Preston
190. Blackpool, Wyre and Fylde
191. Lancaster

Wales
192. Gwynedd
193. Clywd
194. Powys
195. East Dyfed
196. Pembrokeshire
197. West Glamorgan
198. Mid Glamorgan
199. South Glamorgan
200. Gwent

Scotland and Northern Ireland were also affected by the second reorganisation. While the health boards were retained, the districts were abolished in 1983. This left the health services without small geographical divisions for administrative and planning purposes. The loss is likely to be felt most in Scotland, where there are enormous variations in the population and territory covered by each health board (Mair, 1983). While the three island health boards each serve populations of 17,000–31,000, the Greater Glasgow Health Board caters for over 1 million people. Within the Greater Glasgow territory (as in all the others) the administrative structure now fragments directly into management units.

The most recent administrative change has been the creation in England of a Health Services Supervisory Board and an NHS Management Board to take an overview of health services policy and its implementation in the regions and districts (DHSS, 1984d). Previously health authorities had been free to determine their own policies locally: their only real obligation was to balance their budgets. From 1985, the two new boards provide the means for greater central direction.

DISLOCATIONS IN THE SYSTEM

While the administrative structure of the NHS has evolved in an attempt to provide a better service, it still has features which hinder an overall approach to health care planning and the effective co-ordination of services for individual patients. In England, the geographical equivalence of administrative areas established in 1974 has been lost, so the potential for close liaison between family practitioner committees, health authorities and local government authorities is further removed. Dissolving area health authorities did not bring about a large reduction in administrative costs since twice the number of new district authorities with similar duties were created. Of all health districts, 42 per cent have no correspondence with local government territories, 29 per cent match local government territories in multiples and only the remaining 29 per cent have direct correspondence, mostly with local authority districts in metropolitan counties (Population Statistics Division OPCS, 1982). It is accordingly difficult to match the provision of hospital and community health services with the personal social services such as hospital social work, residential and day care for the elderly and handicapped, home helps and services concerned with child abuse, family guidance, alcoholism and drug abuse, all of which are controlled by local authorities. Northern Ireland is in a unique position to integrate these complementary services, yet experience there suggests that a unified administrative system is not enough by itself. Although co-ordination between health and social services in Northern Ireland is said to have improved since they were merged, they still work independently for the most part (Birrell and Williamson, 1983).

Another dislocation is between health authority services and family practitioner services. Family practitioner committees each cover up to six health districts. Their links with health authorities have become more tenuous as they are now spending authorities in their own right, funded directly from the Department of Health and Social Security. The main obstacle to co-ordinated planning is not the mismatch of administrative areas, however, but the autonomy of family practitioner services. Family practitioner committees have an equal number of lay and professional members and they are rarely accused of putting the interests of the public before those of the profession. Their powers, in any case, are limited to enforcing the terms of contracts with individual practitioners. The practitioners themselves remain vigorously independent. As will be discussed later, there is scope for FPCs to be more stringent in approving surgery arrangements, consulting hours and deputising services for general medical practitioners in order to raise the standards of provision in some places. Family practitioner committees make recommendations to the Medical Practices Committee about local shortages of family doctors. This procedure also contains anomalies, to be described in Chapter 3.

Giving the public a greater voice in deciding what health services are needed in local areas might appear to be a way forward towards uniformity in standards. Community health councils (and their equivalents in Scotland and Northern Ireland) were designed to be public watchdogs which can challenge the activities of health authorities if they appear to be against the wishes of the local community. They have the right to receive information on the running of services from the district health authority and the right to make visits of inspection. Community health councils must be consulted by the health authority on any proposals for change in the services in the district, but they have no formal veto and not even a vote, so some critics consider them to be watchdogs without teeth. They do, however, have the power to start an appeal procedure in the special case of hospital closure. If a health authority proposes to close a hospital and the CHC agrees, the closure can take place. If the CHC opposes such a proposal, it can present its case to the Secretary of State, who then makes the final decision.

The style and effectiveness of community health councils vary considerably, according to the interests of the members. Some have been seen as subservient to health service views and others as unduly belligerent. While some have occupied themselves with the details of management, others have adopted political stances. A large number have undertaken surveys of public opinion on particular issues (National Association of Health Authorities, 1983). Heller (1978) provides a detailed example of 'consultation' over a district plan which vividly illustrates his argument that some authorities do not take consultation with the community health council seriously and are more concerned with the semblance than the reality.

At the national level, differences in the provision of health services persist between the constituent parts of the United Kingdom. As will be noted in Chapter 2, there is an argument that health services are supplied more generously in Scotland and Northern Ireland than in England and Wales. A close correspondence is hardly to be expected, since health services are the responsibility of four secretaries of state and four different government departments. In England, health services are administered by the Secretary of State for Social Services through the Department of Health and Social Security (DHSS). In Wales the Secretary of State for Wales is responsible, advised by the Welsh Office. While legislation for England and Wales is usually identical and sponsored by both secretaries of state, health services in Scotland and Northern Ireland are administered independently. The Secretaries of State for Scotland and Northern Ireland direct the health services under their control through the Scottish Home and Health Department and the Department of Health and Social Services in Northern Ireland respectively.

It is evident that the organisational structure of the NHS affords many opportunites for health services to vary in quantity and quality from place to place. However, the National Health Service is more equipped today to recognise and ameliorate geographical variations in services than it was in 1948. Except in inner cities and remote rural areas (whose particular characteristics will be described in later chapters) most geographical differences in service availability appear to be diminishing. It is also worth remembering that the care provided by the National Health Service, impressive though it is, remains small in comparison with the total burden of informal care borne by relatives, friends and neighbours. Organised health services contribute only part of the nation's response to ill health.

2

Regional Variations in Need
and Provision

It is well known that waiting lists for different types of treatment under the National Health Service vary from one part of the country to another. A guide is available that catalogues the relative waiting times in every health district and gives advice on how patients can 'shop around' to find more easily available hospital treatment away from home (College of Health, 1985). There is a clear disparity between the demand for health care and the supply of health services and the severity of the imbalance varies geographically. How, then, can the geographical distribution of health services be made to match the pattern of needs more closely? The question can be approached either by examining the details of where doctors, dentists, nurses and hospitals are to be found (which is the approach of two chapters to follow) or by taking a broader regional view. The broader view is adopted in this chapter. First, an attempt will be made to disentangle the complicated relationships between needs, demands, supply and use. Then the current government policy to match supply to needs at the regional scale will be described and evaluated.

NEEDS, DEMANDS, SUPPLY AND USE

Need and demand

In an ideal world, needs for health care would generate demand which could then be met by supply. Black (1983) has explained why this is not the case in reality. Actual health care needs are distorted by faulty perceptions; there is often little relationship between perceived needs and demand and, finally, supply does not match need or demand.

One of the difficulties is that 'need' cannot be measured in any useful way; indeed, it can hardly even be defined. Various concepts of need have been described, such as 'normative need', 'felt need', 'expressed need' and 'comparative need' (Bradshaw, 1972), but there seems to be agreement that the most useful concept connects need to a medical judgement about the effectiveness of treatment. Brown (1973), for example, offers the definition: 'the existence of a physical or medical condition that would respond to medical or similar treatment'. Seen in this light, people's needs for health care depend upon the state of medical knowledge.

Screening experiments always reveal more illness than is being treated. Brown (1973) reports a screening test in Southwark in 1969 where, out of the first 1,000 cases, 500 were referred to their doctors and only 67 were judged completely fit. For every case of diabetes, rheumatism or epilepsy known to the general practitioner, it is estimated that there is another case undiagnosed. For every treated case of psychiatric illness, bronchitis or high blood pressure there may be five times as many undiscovered (Cooper, 1975). Illness has been compared to an iceberg in that the visible part is only a small proportion of the whole. Statistics on the quantity of treatment delivered by a health service are unlikely to measure the true level of morbidity (illness) in the population. Cooper has estimated that about 20 per cent of illness symptoms result in a visit to the doctor.

Although the actual need for health care is unknown, it is possible to measure perceived needs. The General Household Survey conducted annually in Great Britain has found that about 30 per cent of the population suffer some form of chronic sickness (a long-standing illness, disability or infirmity) and about 12 per cent report short-term restricted activity due to ill health during the two-week period preceding the interview (Office of Population Censuses and Surveys, 1983b). The prevalence of self-assessed ill health varies considerably with personal characteristics. Table 2.1 shows three significant influences on the reporting of long-standing illness. Age has the most dramatic effect, as might be expected, with elderly

21

Table 2.1: Percentage of sample reporting long-standing illness:
Great Britain, 1981

Socio-economic group	0–15	16–44	Age 45–64	65+	Total
MALES					
Professional	16	17	29	28[a]	22
Managerial	15	22	32	51	26
Intermediate	16	20	44	52	27
Skilled manual	16	23	41	55	29
Semi-skilled manual	16	23	44	56	32
Unskilled manual	16	24	48	50	32
All persons	16	22	40	54	28
FEMALES					
Professional	6	19	22	18[a]	18
Managerial	10	21	34	58	26
Intermediate	11	17	40	64	28
Skilled manual	12	22	41	62	28
Semi-skilled manual	15	23	47	63	36
Unskilled manual	11	31	53	63	46
All persons	12	21	41	63	30

Note: a. Based on a sample of 50 or less.

Source: Office of Population Censuses and Surveys, 1983b, p. 142.

people much more likely to report chronic ill health than
younger people. There is also a marked gradient according to
socio-economic group, with people in the professional group
being apparently much healthier than those in the unskilled
manual group. There is also a slight tendency for females to
report more ill health than males. All these trends are repeated in the equivalent figures for acute (short-term) illness.

The geographical implication is that areas which have higher
than average proportions of elderly people, females and
people in the manual socio-economic groups are likely to have
a higher than average prevalence of ill health. Other characteristics of the population may also affect the geographical
distribution of health care needs. Migrants, it is claimed, tend
to be the more active and healthy members of a community,
so areas which have a net gain from migration are likely to
have lower morbidity rates than areas which are losing
migrants (Bentham, 1984). On the other hand, recent migrants
are known to make heavier demands on primary care services
for a short period after moving than established residents

Table 2.2: Use/need ratios by sex and socio-economic group

	Males	Females
Professional	0.23	0.23
Managerial	0.24	0.24
Intermediate	0.20	0.22
Skilled manual	0.18	0.22
Semi-skilled manual	0.20	0.20
Unskilled	0.17	0.19
All	0.19	0.22

Source: DHSS, 1980c, p. 97.

(Cassel, 1977; Freeman, 1978). Immigrants from overseas are a special category with pronounced health care needs (Forest and Sims, 1982).

As a rule, variations in self-reported morbidity are matched by variations in the demand for health services measured by consultation and treatment rates. General practitioner consultation rates, for example, have been found in the General Household Survey to present a profile very similar to that in Table 2.1. Elderly people consult much more frequently than younger people, members of manual occupational groups more than members of professional and managerial groups, and females more than males. There is some debate, however, about whether higher consultation rates among groups which are 'at risk' fully compensate for high morbidity levels, particularly with regard to social class. In the Black Report, 'use/need' ratios were calcuated from General Household Survey data by dividing the number of general practitioner consultations by the number of restricted activity days over a two-week period, for each occupational class. The results appear to indicate that the professional and managerial social groups consult more frequently relative to their needs than the manual groups (Table 2.2).

This conclusion has been challenged by Collins and Klein (1980), who argue that the people reporting needs are not necessarily the people reporting use, so inferences about individuals may be fallacious. They examined the characteristics of 27,154 individuals questioned in the 1974 General Household Survey. These individuals were divided into four health categories according to their reported state of health.

23

In each category, the researchers were interested in the percentage of respondents who had consulted a doctor in the previous two weeks. Disaggregating the results by socio-economic group revealed no consistent bias against the lower groups. The Secretary of State of the day (who was unsympathetic to the Black Report) was quick to draw attention to these results, while others have criticised Collins' and Klein's methodology for its selectivity (Townsend and Davidson, 1982).

The evidence on differences in the uptake of preventive care between social groups is less ambiguous because it is not necessary to attempt to take account of morbidity in this instance. The Black Report presents various data showing that social classes III, IV and V make less demand upon preventive health services than social classes I and II (DHSS, 1980c). Preventive services in which a social class bias has been detected include antenatal services, cervical screening, dental visits and immunisation and vaccination for children. Gatrell's (1985) study of the take-up of whooping cough immunisation in Salford illustrates the resulting geographical pattern: areas with higher proportions of social classes IV and V have lower rates of immunisations.

Some of the differences in the use of health services between different social groups may be due to variations in attitudes, although the attitudes themselves are moulded by external circumstances. The decision to seek medical attention, or to assume a 'sick role' as sociologists term it, depends on a number of factors which differ between social and ethnic groups as well as between people of different ages, sex and marital status (Tuckett, 1976). Low levels of job satisfaction and the need to obtain a medical certificate to cover absence from work, for example, may encourage people in certain occupations to consult their doctor more frequently than otherwise. Other relationships between attitude and social status were revealed by Phillips' (1979b) surveys in West Glamorgan. Three council estate study areas were paired with three areas of owner-occupied housing with a similar age structure and at a similar distance from general practitioner surgeries. Compared with the residents of the higher status areas, the council housing residents were more likely to regard the journey to the doctor's surgery as difficult, less likely to approve of the appointment system (perhaps because they had less access to telephones) and more likely to regard the doctor as an appropriate person to ask about a non-medical problem. While the first two differences in attitude might be expected to reduce the 'use/need' ratio for lower-status residents, the third would tend to increase it. The relationships between morbidity, attitudes and service use for different social classes are therefore quite complex.

Further complication is introduced by the nature and availability of health services. Where no treatment exists, there is no opportunity to translate needs into demands. If a hospital waiting list is thought to be excessively long, general practitioners will tend to avoid adding new patients to it. In this way the demand for hospital treatment responds to the length of the waiting list. New health service facilities will attract demand which was not active previously. People who live in places well supplied with health services generally make more use of them than people in places where health services are less accessible.

Regional distribution of need

Because of the close association between the supply of health services and the demand for them, health service statistics on consultation rates, the number of people treated or the length of waiting lists are very unreliable indicators of need. A measure which is independent of the supply of health services is essential if variations in health from one part of the country to another are to be estimated.

Figures 2.1 and 2.2 show the geographical distribution of two such measures, each with its own drawbacks. Death rates corrected for the differences in age structure between regions (standardised mortality ratios) appear in Figure 2.1. There is a clear gradient from the lowest mortality rates in the south of England to higher rates in the north and Wales. Scotland and Northern Ireland are not included but they, too, have high death rates. High death rates do not necessarily indicate a high need for health services, however. Only a small proportion of health services is devoted to the dying and many incapacitating conditions are not directly life-threatening.

Figure 2.2 gives the regional distribution of the payment of

Figure 2.1: Standardised mortality ratios, 1972

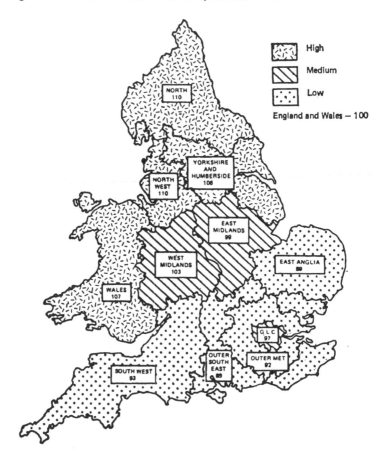

Source: DHSS, 1976, p. 18.

sickness benefit. Women are not included in the statistics mapped because of the large proportion of women who are not insured for sickness benefit. Neither are elderly men or children because they are not employed. The regional variations do not therefore necessarily apply to the whole population, and they may be affected by geographical differences in industrial structure. Social class and the nature of employment are known to influence the likelihood of not working because of sickness. But in spite of these caveats, the regional distribution reinforces that of mortality.

Figure 2.2: Certified spells of incapacity standardised for age, males, 1972

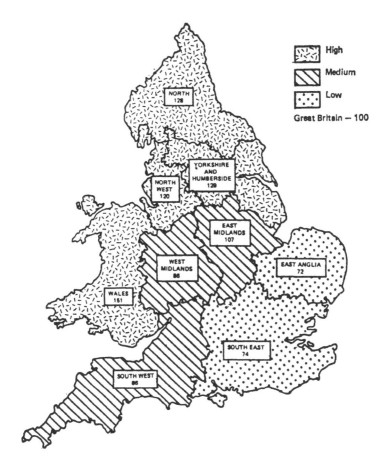

Source: DHSS, 1976, p. 18.

Rates of self-reported·ill health from the General House-hold Survey have the same pattern (DHSS, 1976). Perceived morbidity is not a direct measure of health care needs but it is perhaps as good an indirect measure as is available. Both self-reported morbidity measures, for long-standing illness and short-term restriction of activity, show a gradient from sou-thern England, where the lowest rates are found, to higher values in the north and west.

While each of the measures used is an imperfect indicator of need, taken together they clearly imply that the health of the population in England and Wales deteriorates from south to north. To some extent, the regional differences in mortality and morbidity measures are an artefact of comparing unlike populations. The proportion of the population in social classes IV and V, for example, increases northwards across the country, and people in these social classes are known to suffer greater morbidity than members of other social classes. Variations in social class composition are thought to contribute only a minor amount to regional mortality differences, however. Fox (1977) standardised regional death rates for age and social class and found that substantial regional differences remained. Another factor, whose significance has not been estimated, is the selective migration of relatively healthy people from north to south, which may have the effect of leaving a less healthy population behind.

Most of the explanation is thought to lie in geographical variations in health risks. The climate, for example, is more physically demanding in the north than in the south (winter is the busiest season for health services) and the north has more industrial pollution. The chemical composition of drinking water has a north–south pattern, and this may be a factor affecting the risk of heart disease (Pocock et al., 1980). Dietary habits and the prevalence of smoking also enhance the relative healthiness of the south. For instance, Chilvers (1978) has shown that regional variations in bronchitis can partly be ascribed to geographical differences in cigarette consumption. Both physical and social environments are generally more benign in the south, and the need for health care for a population of given size might be expected to increase moving northwards over the country.

At a more local scale, there are much greater variations in morbidity and mortality rates from place to place (Gardner et al., 1984), but the rates are often based on small population numbers and may therefore be unreliable. Many local concentrations of ill health are likely to be random clusters produced by the vagaries of pure chance, but others could be the result of exposure to noxious elements in the environment, perhaps years or even decades earlier. The debate over elevated levels of cancer incidence in the vicinity of nuclear reprocessing plants (Black, 1984; Craft et al., 1984) illustrates the difficulty of 'proving' a link between ill health and a pollution source, especially to the satisfaction of policy makers and those with vested interests.

Inverse care law

Before the foundation of the National Health Service, the distribution of medical resources was determined primarily by the income level or the rateable capacity of the locality. The number of residents per general practitioner in South Shields, for example, was seven times that of Hampstead. The disparities were recognised to be especially serious because 'under-doctored' districts were usually also poor districts with high rates of sickness and mortality and in special need of a good health service (Political and Economic Planning, 1944). Not only did poor districts have fewer doctors, the doctors they had were likely to base their approach to general practice on a high turnover to guarantee an acceptable income. In other words, the quality of health care was likely to be worst in those places where there was the most urgent need for good health care (Collings, 1950).

Although the removal of geographical disparities in the quantity and quality of health services was one of the objectives of the National Health Service, some differences persisted. Comparing general practitioners in working-class neighbourhoods with general practitioners in middle-class areas, Cartwright (1964) found that those in working-class districts were more likely to have large lists of patients and less likely to have higher qualifications, to hold a hospital appointment, to have access to physiotherapy, contrast-media X-ray facilities or hospital beds, and to visit their patients in hospital. Julian Tudor Hart summarised the problem in *The Lancet*:

> In areas with most sickness and death, general practitioners have more work, larger lists, less hospital support, and inherit more clinically inefficient traditions of consultation than in the healthiest areas; and hospital doctors shoulder heavier case-loads with less staff and equipment, more obsolete buildings, and suffer recurrent crises in the availability of beds and replacement staff. These trends can be summed up as the inverse care law: that the availability of

good medical care tends to vary inversely with the need of the population served. (Hart, 1971, p. 412)

The strongest evidence in support of Hart's 'inverse care law' is of deficiencies in primary care in socially deprived inner-city areas (see Chapter 7). In the most deprived areas a sub-standard primary health service reinforces the cycle of poverty in which low incomes, unemployment and poor housing conditions combine with poor educational opportunities to produce successive generations of people with low levels of skills, low aspirations and poor health. Not only is social deprivation associated with low levels of health, it also tends to create additional pressures on health services. People who live in sub-standard housing or accommodation lacking the basic amenities are likely to need longer stays in hospital before being discharged. The same is true of people who are not supported at home by a social network of family and neighbours. People with poor levels of education and social skills make little use of preventive services but may use the expensive facilities of the accident and emergency service for primary health care needs at times of crisis. In the areas where need is greatest, services are least able to cope with the pressure (London Health Planning Consortium Study Group, 1981).

The inverse care law has also been detected in operation at the regional level. Coates and Rawstron (1971) found pronounced regional discrepancies in the supply of general practitioners and hospital beds, with the worst provision associated with the poorest regions. Noyce et al. (1974) investigated regional differences in expenditure on community health services and found it to be positively associated with the proportion of professional and managerial workers in the population and inversely related to the proportion of manual workers. Furthermore, low spending on community health services was associated with low spending on hospitals and vice versa, so there was no likelihood of generous provision in one service compensating for deficiencies in another. Heller's (1978) study of health districts in East Anglia produced the same conclusion: districts with poor hospital services had poor community services, and, to make matters worse, low levels of social services provided by the local authority.

Figure 2.3: Relationship between hospital use and provision of beds for acute cases in health regions, 1977

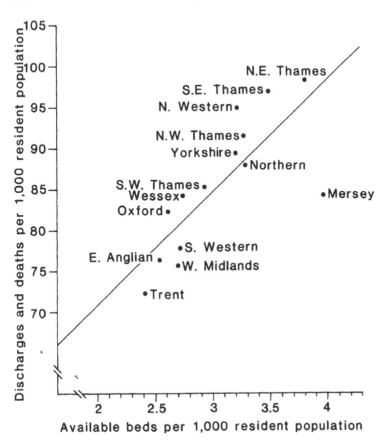

Source: London Health Planning Consortium, 1979, p. 24.

Supply and use

If health services are in greatest supply where they are least needed, there might be excess capacity in the most favourable areas. This is not the case. Whatever the background level of need, the services provided are used. Demand adjusts itself to supply. West and Lowe (1976) compared the need for child health services (measured by several indices of birth and death rates) in hospital regions with the provision of child health services and their use. They found generally negative correlations between need and provision, indicating that the regions with most need had least services. The use of child health services was also negatively associated with need, but use was directly related to provision.

Cullis *et al.* (1981) examined variations in general hospital bed use rates within and between regions. Their conclusion was that bed use rates were highly correlated with bed supply rates and only very weakly associated with variations in mortality rates. Figure 2.3 tells the same story. The discharges and deaths per thousand resident population are plotted against the number of hospital beds per thousand resident population for each health region. There is a positive relationship between the two measures. When the same exercise was repeated for hospital catchment areas in the four Thames regions, the relationship was even stronger, with a correlation coefficient of 0.85 (London Health Planning Consortium, 1979). This demonstrates that the use of hospital beds is dependent on the number of beds available: areas with large numbers of beds have large numbers of people making use of them. It also shows the dangers of interpreting health service treatment statistics. Hospital utilisation statistics do not distinguish deaths from discharges, so they do not measure successful treatment (maternity statistics are an exception); neither do they measure the morbidity of a population. Variations in the number of people receiving health service treatment from one place to another reflect differences in the provision of services, not need.

Dental care: needs, supply and use

The same relationships between needs, supply and use have been demonstrated for dental services, for which tooth decay provides an unusually good measure of need. Payments to dentists are based on a detailed record of services rendered which is held in machine-readable form by the central Dental Estimates Board. These data were used by Ashford (1978) in a study of the regional variations in dental care in England and Wales. For 1974, Ashford found that the number of dentists per 10,000 population varied considerably from 4.18 in NW Thames and 3.64 in SW Thames down to 1.76 in the Northern

Table 2.3: Percentage of 14-year-olds with some permanent teeth extracted for decay, by dentist attendance frequency and region

Dental attendance pattern	The North	Midlands and East Anglia	Wales and the South West	London and the South East	Wales alone
Regular check-ups	36	31	26	18	40
Occasional check-ups	33	42	40	11	43
Only when trouble	41	40	49	28	63

Source: Todd, 1975, p. 194.

Region and 1.73 in the Trent Region. London and the south had the most favourable supply of dentists. Comparing the annual number of dental treatments per person in the various regions with the supply of dentists revealed a positive linear relationship. The regions with the greatest supply of dentists had the highest per capita use of the general dental service. This association between supply and use was most pronounced amongst children and young people: the very section of the population for whom good dental care is most critical.

As with other health services, regional variations reflect social differences. Todd's (1975) study of children's dental health in England and Wales showed a clear social class gradient in terms of parental knowledge of dental health care, parental attitudes towards treatment, parental expectations concerning their children's teeth, children's frequency of attendance for dental examination and treatment, and the physical conditions of children's teeth. It also revealed geographical trends. In the regional classification used by Todd, London and the South East had the most favourable ratio of dentists to population, followed by the South West, the North, the Midlands and East Anglia and, finally, Wales. Attendance rates followed the same pattern, with the highest attendances where they were most dentists, but tooth decay had the reverse distribution. Regions with the most tooth decay had the least dentists. Part of the reason might be found in attendance variations, but the regional variations in tooth quality were still evident when children with different attendance habits were considered separately (Table 2.3).

Since areas with relatively few dentists are often those with larger than average proportions of their population in social classes IV and V, it is possible that variations in dental attendance between areas could be the result of social class differences in attitude towards treatment rather than differences in the supply of dentists. O'Mullane and Robinson (1977) demonstrated that this was not the case. These authors examined 14-year-old school children in two towns, one of which had a more favourable dentist to population ratio than the other. They found that children from different social classes had similar high levels of treatment in the better provided town. In the town with relatively few dentists, treatment levels were lower but there was a marked discrepancy between the social classes: children from social classes IV and V were much more disadvantaged than those from classes I and II. In other words, use of the general dental service was related to supply, but the strength of the relationship varied between social classes.

Social class differences in the uptake of treatment are partly due to variations in attitude, but they may also be prompted by variations in physical accessibility. Bradley, Kirby and Taylor (1976) found that the proportion of all children examined in 23 primary schools in the Newcastle area who needed dental treatment was negatively correlated with measures of access to dentists' surgeries. The schools whose pupils had the highest needs for dental care were those furthest away from dental facilities. Working-class areas were most disadvantaged (Taylor and Carmichael, 1980).

Rationing health services

The National Health Service faces ever increasing levels of demand as the iceberg of morbidity emerges. The availability of services sharpens the perception of needs in the population and people's tolerance of ill health is less than it was. The service has taken up some of the increased demand firstly by expanding, which occurred up until the mid-1970s, and secondly by increasing efficiency. Perhaps the single most important contributor to higher efficiency has been the policy of using hospital beds more intensively by discharging patients earlier than formerly. A third strategy has been to prevent ill health occurring. Some services are provided for the sole purpose of prevention, such as ante-natal clinics and immunisation. The health service also promotes information campaigns against cigarette smoking, for instance, and acts as a lobby for campaigns such as that for the compulsory wearing of seat belts. The proportion of resources devoted to prevention is small, however, and many argue that the National Health Service should devote much more of its energy to this type of work.

The fourth, and final, strategy which has been used in response to the pressure of demand is the rationing of treatment. In some cases this takes the form of persuading the potential patient that he or she should manage without help: prescription charges fall into this category, as they are partly intended to curb demand. Propaganda campaigns have been mounted to discourage people from consulting their general practitioner unnecessarily. Another rationing device is the waiting list: people may die during the waiting period or circumstances may change so that treatment is no longer demanded. The length of the waiting list is sometimes used as a measure of the inefficiency of the service, but this interpretation is misleading. Waiting lists are expressions of demand, and when there is no realistic hope of treatment the demand shrinks away. Other rationing mechanisms include decisions within the health service on what services to emphasise and which to assign a lower priority. The acute illnesses of younger people, for example, are given much more priority than the chronic conditions of the elderly. The accessibility of health services is yet another rationing mechanism whose effect is to exclude some people from the benefits of the service. According to the inverse care law, the people with the most pressing needs are often the ones to be excluded.

While some forms of rationing are unavoidable, others seem obviously inequitable. Geographical variations in service provision are particularly unacceptable in a service whose aim is to provide equal access for all. At the root of the problem is the supply of finance, which in the past has not been sufficiently directed to the areas of greatest need. Shifting the geographical pattern of finance is not an easy task, however, especially as the criteria on which to base the shift are far from clear.

GEOGRAPHICAL REDISTRIBUTION

Criteria for resource allocation

Policies to make the distribution of health services more equitable have concentrated on the regional pattern of hospital provision. Hospitals have always been the single most expensive part of the National Health Service and, because of the way they are organised, they are more amenable to central control than the primary health care services (which are to be discussed in the next chapter).

Before the National Health Service was established, hospitals were built and run either by private benefactors or by local authorities. The wealthier areas generally had better hospital services than the less prosperous areas. After 1948, when health services first began to be financed on a national basis, the existing buildings and services had to be maintained, so regional hospital boards were allocated funds in proportion to their commitments. As a result, the regions which were better off in terms of hospitals received proportionally more money than the poorer regions and the imbalances continued. In 1971, for example, expenditure per head on hospitals in the Sheffield region was only 55 per cent of that in the South West (National Association of Health Authorities, 1983). The good intentions of the NHS to provide an equitable service across the country were not being met.

The obvious solution was to allocate to each region the same amount of money per head of population, but there were several arguments against this. First, to avoid the problem of subtracting services from the better endowed regions in order to provide more elsewhere, the equalisation process should take place gradually through differential growth. There was no suggestion, after all, that the more fortunate regions should lose health services because they had too many, but a case could be made for services in these regions to grow less quickly than services elsewhere. Secondly, population totals alone are a poor measure of need for health resources. A region with a high proportion of elderly people, for example, probably deserves more money per capita for services than a region with a younger population. Some parts of the country are healthier than others and so presumably should be given less money per capita than unhealthy areas. It would clearly be wrong, moreover, to ignore geographical variations in costs. Some regions have larger commitments to expensive clinical teaching facilities than others, some regions with specialised services regularly import patients from outside their boundaries while other regions export. There are also extra costs involved in providing services in London.

What was needed was a method of dividing money between the regions in a way that would reflect these considerations. The first attempt was in 1971, when the allocation was made on the basis of both existing provision (numbers of beds and numbers of cases treated) and need (population weighted for age and sex composition). The present policy based entirely on estimates of need was introduced by the Resource Allocation Working Party in 1976.

The RAWP method

The Resource Allocation Working Party (RAWP) was appointed:

> To review the arrangements for distributing NHS capital and revenue to RHAs, AHAs and Districts respectively with a view to establishing a method of securing, as soon as practicable, a pattern of distribution responsive objectively, equitably and efficiently to relative need and to make recommendations. (DHSS, 1976)

Within these terms of reference, the working party interpreted its task as that of devising a formula to divide the financial cake between the regions for capital expenditure ('one-off' building and major equipment spending), revenue expenditure (for day-to-day running costs) and teaching expenditure. Deciding how best to spend the money within these headings was left to health authorities. The objective set out in the working party's report was that 'there would eventually be equal opportunity of access to health care for people at equal risk' (DHSS, 1976, p. 7). Towards this end, the working party attempted to identify the best measures of variations in need between different parts of the country. Its recommendations were adopted forthwith as government policy and have been applied in all allocations since 1976 with slight adjustments after a review in 1980 (DHSS, 1980b). They may be summarised as follows.

1. Size of population.

The mid-year estimate of a region's population is the basic measure of the overall level of need.

2. Age and sex composition.

Population size is weighted by a number which makes allowance for age and sex distribution. For example, a region with higher than average proportions of women and elderly people (and therefore presumed to have higher than average health needs per capita) receives a positive adjustment.

The weighting applied is calculated from the distribution of the population in male and female ten-year age bands and the health service usage rates of each age/sex category extracted from national statistics.

3. Mortality rates.

To take account of geographical differences in health as far as possible, population size is next weighted by the standardised mortality ratios (SMRs) of the region. The standardised mortality ratio is the ratio of actual deaths in a region to the number expected on the basis of national rates applied to the region's particular age and sex distribution. Regions with age and sex adjusted death rates higher than average receive a positive weighting, and *vice versa*. These calculations are done for all the main causes of death (each with its own SMR) and the results are amalgamated by giving each cause of death a weight proportional to its hospital bed usage rate.

The effects of weighting populations according to age and sex composition and standardised mortality ratios are shown in Figure 2.4. According to the working party's calculations, the Oxford and West Midlands Regions had the most favourable population structure and SE Thames and the South Western Region the most demanding age and sex composition in terms of health services. Adding the variations in mortality rates changes the rank order of regions substantially. Oxford and the West Midlands still receive the largest negative adjustments, with the addition of NW Thames, but the highest positive weightings are given to the Yorkshire and North

Figure 2.4: The effects of weighting regional populations by age, sex and standardised mortality ratios

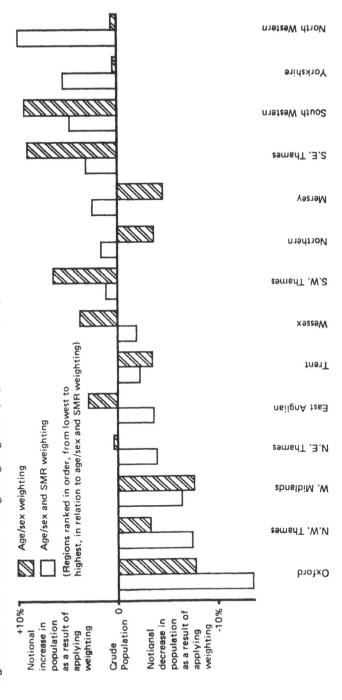

Source: DHSS, 1976, p. 20.

Western Regions, where standardised mortality ratios are particularly high.

4. *Cross boundary flows.*

Adjustments are made for the movement of patients across health authority boundaries in calculations based on average costs for different specialties.

5. *Cost weighting.*

The effect of the London weighting on pay is taken into account in the four Thames regional health authorities.

6. *Service increment for teaching.*

Regions are allocated extra funds according to their numbers of medical and dental undergraduates. The amounts are based on the average costs of training medical and dental students.

7. *Capital stock.*

Capital allocation, which is treated separately from revenue allocation, has an additional adjustment for the value of existing capital stock (mostly buildings) and also its age and condition.

Steps 2 to 5 are carried out separately for the six main groups of services provided by health authorities. The groups of services are those for non-psychiatric in-patients, day and out-patients, mental illness in-patients, mental handicap in-patients, community and ambulance services. For each group there are slight variations in the method. In the case of mental handicap and mental illness in-patient services, for example, it is not appropriate to adjust for mortality rates, since mortality rates are very poor indicators of the prevalence of mental disorders. The final stage is to combine the population weightings for all six service groups to produce an overall weighted population for each region, which is the index of that region's relative 'need'.

Moving towards the target

The end result of the whole process is two 'target' allocations for each region − one for revenue and the other for capital −

Figure 2.5: Regions' distance from RAWP revenue targets, 1977–86

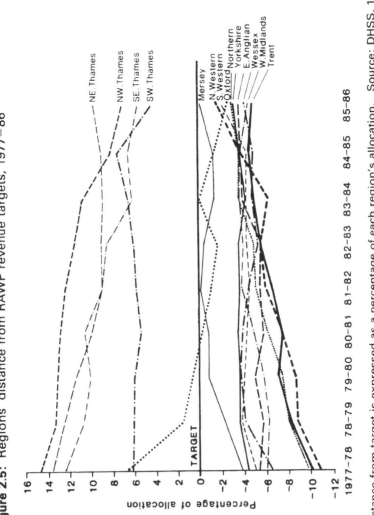

NE Thames
NW.Thames
SE.Thames
SW.Thames

Mersey
N.Western
S.Western
Oxford
Northern
Yorkshire
E.Anglian
Wessex
W.Midlands
Trent

TARGET

Percentage of allocation

16
14
12
10
8
6
4
2
0
−2
−4
−6
−8
−10
−12

1977–78 78–79 79–80 80–81 81–82 82–83 83–84 84–85 85–86

Distance from target is expressed as a percentage of each region's allocation.   Source: DHSS, 1985b, p. 3.

which indicate how the available money should be divided in order to achieve a fair distribution which matches the needs of regions. Revenue expenditure is approximately fifteen times that of capital expenditure.

Comparisons between the regional target allocations and the shares of resources they had previously been getting revealed considerable discrepancies, with a geographical bias in favour of the south-east. The four Thames regions and Oxford were shown to be getting proportionately more revenue than they should under the RAWP method. The remaining English regions were below target, with the North Western, Trent and Northern regions in the worst relative position. The distribution of capital stock compared with the target capital allocation also showed differences with a similar geographical pattern, except that the Mersey Region was in a particularly favourable position due to its inheritance of a strong tradition of philanthropic foundations and municipal hospitals.

Arguing that immediate adjustments to the RAWP targets would be too drastic, the working party recommended a gradual movement of allocations towards the target. Figure 2.5 illustrates how this has taken place, with a steady reduction in regional differences. By 1984–5 the extremes between actual revenue allocations and targets were West Midlands and Trent (5 per cent below target) and NE Thames (9 per cent above target).

For 1985–6 the overall growth in revenue expenditure was 5.5 per cent above the previous year and this was distributed so that the region furthest below target was given a growth rate of 6.8 per cent and the regions furthest above target grew by 4.2 per cent (DHSS, 1984c). These figures are somewhat misleading because the 'growth' in some regions is in fact a decrease in funding in real terms, once the effects of inflation and an ageing population are taken into account. The fact that the proportion of elderly people in the population is increasing means that real resources must increase by about 1 per cent per year to maintain the same level of service availability (Institute of Health Service Administrators, 1984). The Thames authorities and the Oxford Region have some justification for arguing that the RAWP policy has led to cuts in services which were never shown to be over-provided.

Allocations to districts

The working party was primarily concerned about making the allocations of resources to regions more equitable, yet it was clear that the existing distribution of resources was even more variable at the sub-regional level (Rickard, 1976b). The RAWP report recognised that the reallocation would achieve its objective only if carried through to the district level, where the services are actually provided. If regional authorities used different criteria when allocating funds to districts there was a danger of widening the disparities between districts, so it was proposed that regional authorities should distribute resources within the region by applying a similar formula. Some difficulties were acknowledged. One problem is that district populations are much smaller than regional populations and consequently the statistics of health service use and mortality are much less reliable at district level, being subject to relatively large random fluctuations. Even averaging over twenty years would not necessarily produce the true average for condition-specific SMRs in districts (Geary, 1977). Another difficulty is that patient flows across boundaries are much more significant at district than at regional level, so the estimation of the costs and savings involved has a large impact on district totals (Senn and Shaw, 1978). The working party therefore warned against a rigid interpretation of the method within regions, while commending its value in giving an objective yardstick of relative disparity between districts.

In practice, the methods used by regional health authorities to identify the relative needs of health districts have varied. Some have followed the RAWP proposals, others have modified the RAWP method (for instance by using overall SMRs rather than condition-specific SMRs, to avoid the problem of small data bases) and others have used totally different methods. At the local level, the problem of determining appropriate funding for regional specialties and centres of excellence has been particularly troublesome (National Association of Health Authorities, 1983).

SHARE in Scotland

The Scottish equivalent of the RAWP report appeared a year later under the title Scottish Health Authorities Revenue Equalisation, or SHARE (Scottish Home and Health Department, 1977). Although the method recommended by the Scottish working party was broadly similar to that of RAWP, there were differences in detail because of the much smaller size of health boards compared with English health regions. The population of the 15 health boards ranges from less than 20,000 to about 1 million, whereas the smallest English region contains almost 2 million people, so by comparison the Scottish boards are much less self sufficient and cross-boundary flows of patients are more significant. The most striking feature of cross-boundary flows at the time of the SHARE report was the huge import of patients into the Greater Glasgow Health Board. Lothian, Lanarkshire and Tayside were also large importers of patients, but not on the same scale. Lanarkshire, Argyll and Clyde, Greater Glasgow and Fife were the largest exporters. Mindful that the costing of cross-boundary flows would have a large effect on the final measures of relative need, the Scottish working party classified hospitals into 45 types and estimated the costs of cross-boundary flows according to the average costs of the type of hospital in which patients were received.

The smaller size of Scottish health board populations prompted another departure from the RAWP model. Because they were based on small numbers of deaths, the cause-specific death rates were felt to be unreliable. The standardised mortality ratios for all causes of death were used instead (and only for the population aged under 65).

A third departure was an adjustment made for population density when calculating allocations for community health services. It was expected that a higher provision of district nurses and health visitors would be required to make allowance for higher proportions of time spent travelling in the sparsely populated parts of Scotland. A sparsity factor was calculated for each health board, based on a crude index of the average distance of patients from their general practitioners. Values ranged from 0.040 and 0.320 for Greater Glasgow and Lothian Health Boards respectively to 3.131 for Dumfries and Galloway, 3.256 for Borders, 4.892 for Highland and 8.065 for

Figure 2.6: The results of SHARE: Scottish health boards' allocations

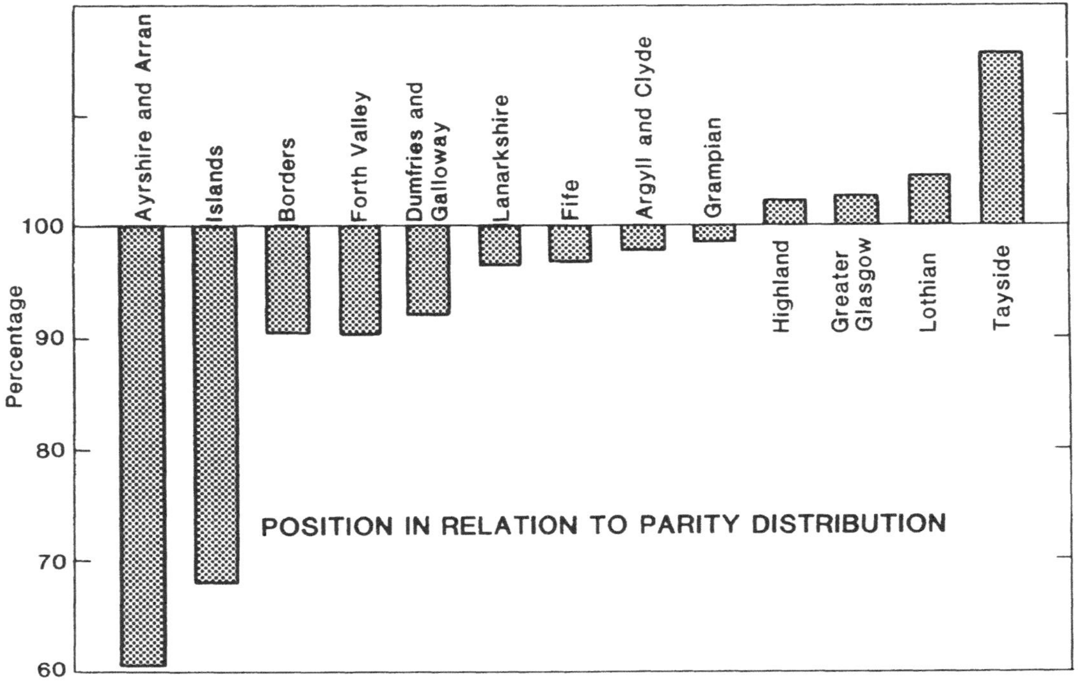

POSITION IN RELATION TO PARITY DISTRIBUTION

Source: Scottish Home and Health Department, 1977, p. 42.

the Islands Health Boards. These sparsity factors were used to weight 30 per cent of the allocations for district nurses and health visitors (since community nurses were expected to spend 30 per cent of their time in travelling). The overall effect of the age, SMR and sparsity weighting was to give the Islands Health Boards a weighted population twice that of their crude population for community services. The Highland Board also gained substantially from this procedure, with a ratio of 1.5 weighted to crude population for the calculation of community services allocations.

The results of the SHARE exercise are illustrated in Figure 2.6. The differences between the SHARE targets (the parity distribution) and the actual allocations in 1975−6 were even greater than those revealed in England. Ayrshire and Arran and the Islands Health Boards (a combination of the Western Isles, Orkney and Shetland Health Boards) were shown to be 30−40 per cent short of their equitable allocation as determined by SHARE. Tayside, Lothian and Greater Glasgow Health Boards, containing the bulk of Scotland's hospital services, together with the Highland Board, were all over target. As in England, the recommendation was that allocations should move towards parity gradually over a period of years. Although the largest proportional reductions at the time were indicated for Tayside, the working party drew attention to the special problems of the Glasgow Health Board, whose allocation is expected to fall markedly during the remainder of the century because of population movements out of the city and a decline in the present high level of patient imports.

Wales and Northern Ireland

Wales and Northern Ireland have their own methods of distributing resources to district health authorities and health and social services boards respectively, also modelled on the RAWP approach. For Wales, the policy was established in the SCRAW report (Welsh Office, 1977) and for Northern Ireland, the PARR report (Department of Health and Social Services, 1978). Like SHARE, both SCRAW and PARR are concerned with revenue and not capital. They also use overall standardised mortality ratios as weighting factors rather than condition-specific ratios because of the small number of deaths involved. Both use sparsity factors for ambulance services (road miles per capita in Wales and the ratio of local to national average miles per patient carried in Northern Ireland). More detailed comparisons are made in Maynard and Ludbrook (1983).

The results of the SCRAW exercise in Wales revealed that Dyfed was the district furthest below target (−9.71 per cent), with West Glamorgan, Mid Glamorgan and Gwynedd also below target. The four remaining districts were above target, with South Glamorgan (+8.62 per cent) in the most favourable position. In Northern Ireland, the differences between boards were smaller. Calculations showed that the Northern Board's allocation should be increased by 4.3 per cent and the other three allocations reduced by small proportions.

Technical criticisms of the RAWP method

Considerable doubt has been expressed on whether mortality is a good indicator of the need for health care, especially for non-fatal conditions like arthritis which make heavy demands on health services. Some fatal conditions are related not directly but inversely with demands on the health service (as mortality increases, demand falls). Infant mortality is the extreme example. It is not even clear that the areas with high mortality rates also have high morbidity. Several researchers were quick to point out that patterns of mortality and morbidity do not necessarily match (Ferrar et al., 1977; Forster, 1977). The working party had anticipated these criticisms, and justified its use of mortality statistics with the argument that they are simply the best measure of need currently available. The advantages of mortality statistics are that they are of high quality, sufficiently detailed to permit comparisons by place of residence and medical condition, they cover the whole population and they are independent of service supply. The self-reported morbidity measures recommended by Curtis (1984) may be a better substitute, especially for local areas in future, but lack geographical coverage at present.

The choice of the standardised mortality ratio to be used as the measure of mortality has also been criticised. The SMR is heavily affected by deaths among elderly people, which might obscure more significant and reliable mortality variations

Table 2.4: National comparisons of resources and standardised mortality ratios (SMRs)

	Expenditure per capita in 1982/3 (£)				SMR (1983) (UK = 100)	
	Hospital Services	Community Health Services	Family Practitioner Services	Total[a]	Male	Female
England	142	16	55	244	98	98
Wales	150	17	62	263	106	102
Scotland	192	18	57	309	112	111
N. Ireland	189	19	64	313	109	108

Note: a. Includes capital expenditure and other services not shown separately.

Source: Derived from Central Statistical Office, 1985a, pp. 57, 68.

among younger people. Other measures could be used instead, such as age-specific death rates (Barr and Logan, 1977) or crude death rates (Cochrane, 1976; Ferrar, 1977). This controversy was dampened by Palmer *et al.* (1979), who inserted several different mortality indices into the RAWP formula and found that the SMR was among the measures which produced the fewest regional differences, so its effect is safely conservative. In a situation of such uncertainty it is probably better to risk underestimating regional differences than overestimating them.

One aspect of the mortality weighting method has attracted a more fundamental comment. To make sure that common conditions have more influence on the overall result than rare ones, condition-specific SMRs are weighted by existing condition-specific bed usage rates in RAWP (although not in the other methods). This procedure resembles previous methods of resource allocation in that it tends to perpetuate existing patterns of use, and does nothing to promote the health service's expressed priorities in providing more care for relatively neglected conditions or groups of people (Buxton and Klein, 1978).

Cross-boundary flows are difficult to deal with, especially between districts. Patients receiving hospital care outside their own district or region tend to be those with complex conditions requiring specialised help not available nearer home. Their treatment is likely to be more expensive than average. The use of average costs will therefore discriminate against net importers of patients, like the four London regions. NW Thames, for example, imports about 15 per cent of its patients and exports only about 5 per cent (Buxton and Klein, 1978). At the district level, the method could work against the districts with teaching hospitals and in favour of those without.

The allowance specifically made for teaching hospitals has also been controversial. It takes into account the costs of training medical and dental students but not physiotherapists, nurses and other staff. Neither does it allow for the costs of conducting research, pioneering new techniques and demonstrating high standards of care. Acknowledged 'centres of excellence', whose influence in the long term benefits the entire hospital system, receive no special funding. The London authorities in particular claim the injustice of this omission.

A final technical criticism is that RAWP, SHARE, SCRAW and PARR do nothing to address the question of equity of expenditure on health services between the four parts of the United Kingdom. Table 2.4 shows that expenditure per capita is highest in Northern Ireland and Scotland. Taking account of variations in mortality rates would reduce the differences somewhat. The Royal Commission recommended that there should be an explicit formula for the distribution of health service funds to the four parts of the United Kingdom, but political considerations are likely to stand in the way. Application of the RAWP formula to the constituent parts of the United Kingdom would shift substantial resources from Scotland and Northern Ireland to England and Wales (Maynard and Ludbrook, 1980).

Fundamental criticisms of RAWP

A serious deficiency in the redistribution process is that it excludes the family practitioner services, which are provided by independent contractors and are not subject to cash limits in the same way as other health services. In the four-year period up to 1984, family practitioner services increased their share of central government expenditure on health from 22.1 per cent to 23.4 per cent (DHSS, 1985a), largely through uncontrolled spending on drugs. Family practitioners are neither administered nor financed by health authorities, so they remain outside the RAWP process. The national distribution of family doctors, dentists, opticians and pharmacists is far from being even, and the present method of funding them simply funnels more money into the areas of greatest supply.

The allocation methods have also been criticised for making no special allowance for areas of multiple deprivation, where large concentrations of vulnerable social groups live in poor environmental conditions (Avery-Jones, 1976). The SMR adjustment goes some way to compensate for the deleterious effects of social class (Townsend, 1981), but perhaps not far enough. The RAWP method makes no attempt to make up for deficiencies in housing, employment and environmental services. No account is taken of geographical variations in social service provision, even where there is obvious overlap, as in the case of residential care for the elderly. The treatment of the health service as an independent entity whose needs are assessed separately from those of other services is most likely to disadvantage inner-city communities, whose demands on the health service are increased by shortages in other benefits.

Demographic trends have also put inner-city areas at a disadvantage in the RAWP approach. Application of the RAWP formula in large conurbations, where the outer suburbs are growing and the inner city is declining in population, shows the inner-city areas to be above-target and likely to have services reduced, in spite of the poor social conditions there. The London authorities in particular, with little or no growth money to spend, have had to subtract services from some places in order to make good deficiencies in others.

An example cited by the Radical Statistics Health Group (1977) illustrates one effect of the RAWP policy at the local level. It concerns the City and Hackney District in NE Thames Region, an inner-city area with the attendant problems of social deprivation, which had to make savings. The district contained two very different hospitals, the Metropolitan Hospital, whose patients were all from within the district, and St Bartholomew's Hospital, with an international reputation and only 13 per cent local patients. After considering the issues, the health authority decided to maintain St Bartholomew's at its existing level and make the savings required by closing the Metropolitan Hospital. There was obviously a case for this decision, but it could be claimed that the losers were the local people, even though their needs are demonstrably high.

Yudkin (1978) provides a further illustration, from Tower Hamlets District, another socially deprived area in the NW Thames Region, the region which has lost most under the redistribution policy. In Tower Hamlets substantial savings were planned by reducing the number of beds in three small hospitals and closing a fourth. The beds lost were used primarily for social admissions of local people. Meanwhile, plans for a new clinical block at the London Hospital with running costs of £500,000 a year went ahead. Yudkin concluded that the reallocation of resources may have little effect on the funding of teaching hospitals, but may drastically affect services for the most socially deprived.

In the competition for funds between local facilities and centres of excellence, the groups with most professional and political influence are clearly in the best starting position. The allocation of resources below district level is a matter of choosing priorities between the different branches of the service. Districts with the same financial resources may use them differently, and variations in resources are not necessarily accurate indicators of variations in the availability of health care. Availability depends on the mix of services and the location and efficiency of these services: matters which the RAWP method does not address.

The RAWP approach sets out to achieve a more even distribution of resources down to the district level. The problem does not stop there, however, for the most marked geographical differences in access to health services occur locally, within districts. Absolute geographical equity is an impossibility, since population and services cannot be in the same place, but ensuring a fairer geography of health care requires matching services to the distribution of population. Services provided for sparse rural populations cost more per capita than services for dense urban populations. In Scotland, an allowance is made for population density, but not in England.

With the objective of providing 'equal opportunity of access to health care for people at equal risk' the strategy of the Resource Allocation Working Party, and subsequently of the government, has been to redistribute resources geographically down to the districts. Other forms of redistribution were not adopted. These include redistribution according to social disadvantage and redistribution between the branches of the health services, particularly from acute services to chronic and preventive services. These alternative forms of redistribution have a greater chance of achieving the objective of equal opportunity of access than the geographical method, it might be argued. A radical view is that the geographical method may even hinder change towards greater social equity.

Perhaps the most radical stance is that RAWP is irrelevant to the real issues. RAWP makes no contribution to the debate about the proper level of health care; it is simply a rationing mechanism to distribute the amount made available. The exercise can be interpreted as a smokescreen, generating obscure controversies which hide the fundamental problem. Most health service workers would say that insufficient money for health services is the real problem. Some, however, have argued that increased expenditure on health services would not necessarily bring a concomitant improvement in the population's health. The Royal Commission itself produced evidence to show that there is no relationship between expenditure on health services and perinatal mortality rates in countries with well developed economies (Royal Commission on the NHS, 1979a). Better health services do not appear to reduce death rates significantly (Cochrane *et al.*, 1978) and their effects on the prevalence of illness are comparatively modest (McKeown, 1979). Making access to health services more uniform is not the same as giving people equal opportunities for health.

Local details

Local variations in service provision cannot be removed while regional discrepancies remain, but the achievement of regional uniformity will not guarantee equal access to services for individual consumers. Regional policy is a blunt instrument. What matters to the patient is the care provided by a particular doctor or a particular hospital. The next two chapters accordingly turn to the details of specific health services.

3

The Location of Primary Care

The primary health services are the point of first contact for most National Health Service consumers. They consist of the family practitioner services (general medical and dental practitioners, ophthalmic specialists and pharmacists) and the community health services (such as community nursing, health clinics and the school health service). As was described in Chapter 1, the family practitioner services are much more independent than any other branch of the NHS. All family practitioners make their own consulting arrangements, employ any staff they consider necessary and usually provide their own premises, although increasing numbers rent space in health centres owned by health authorities. The community health services, in contrast, are managed by health authorities. Although they are organised separately, the effectiveness of each group of services is to some extent dependent on the other. In this chapter, the factors influencing the geographical distribution of primary health services are discussed with a view to identifying gaps in coverage and possible ways to improve the pattern of care provided. We start with a key figure: the family doctor.

GENERAL MEDICAL PRACTITIONERS

General practitioners' lists

Each doctor who sets up in general medical practice by contracting his or her services to a family practitioner committee

Figure 3.1: Surgery attendance patterns of West Cross residents in the Greater Swansea area

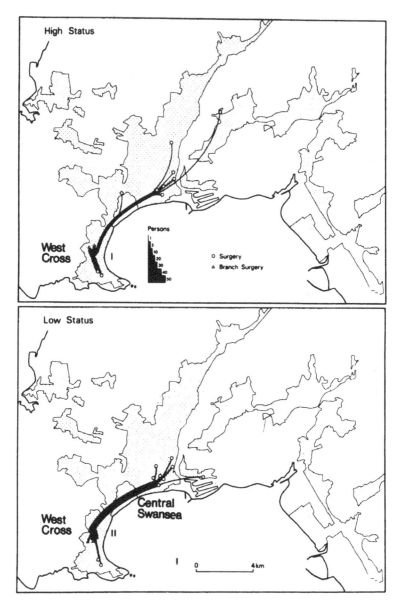

Source: Phillips, 1981, p. 125.

or health board has a personal 'list' of patients. These are the people registered with that particular general practitioner. People have a right to choose their own general practitioner, but general practitioners are not obliged to accept any application to join their list. In general, people choose a doctor whose surgery is near. Cartwright and Anderson's survey (1981) found that about half the people questioned said their doctor's practice was the one nearest to them. As many as 45 per cent said either they themselves or their doctor had moved since they registered with him or her. Moving but keeping the same doctor is the main reason a substantial proportion of people are not registered with the nearest general practice. Phillips (1979a), in a study of the Greater Swansea area, found that people who had always lived locally or who had moved into the area from elsewhere were likely to be registered with the nearest general practice, whereas people who had moved a few kilometres within the area were likely to have retained their previous general practitioner. A common pattern was for families which had previously lived in central Swansea to retain their former general practitioner after moving to one of the study sites on the outskirts of the city (Figure 3.1). The two adjacent study sites in Figure 3.1 consist of owner-occupied housing ('high status') and a council estate ('low status'). In the council estate, 21 of the 52 people interviewed had originally lived in central Swansea and 17 of them had not changed their practice since moving. This suggests that, within certain limits, distance to a surgery is a secondary consideration to many people who wish to maintain contact with their established general practitioner. Of course, doctors may actively discourage such contacts because of the suspicion that they bring an increased demand for home visits (Hopkins *et al.*, 1968). It is in the interests of patients and general practitioners (who are obliged to make home visits when the patient's medical condition does not allow travel to the surgery) for the distance between home and surgery to be small. Nevertheless, general practitioners are in competition with one another for patients and the catchment areas of their practices overlap considerably, especially in densely populated urban areas.

The average list size for general practitioners in the United Kingdom was 2,062 in 1983 (Central Statistical Office, 1985a), although substantial variations occur. General practitioners who work in sparsely populated areas, those who undertake large amounts of private work and those who are semi-retired usually have smaller lists than average and those working in unattractive areas with relatively few other doctors have larger lists. Almost invariably general practitioners have fewer real patients than their list size would indicate. The difference between the number of patients whose names appear on the list of a practice and the actual number of patients on the list who still reside in the area is known as 'inflation'. An inflation of 20 per cent, indicating that only 80 per cent of the names on the list are 'real' patients, is not uncommon (Morrell *et al.*, 1970). While patient delay in registering with a doctor causes under-registration, the number of over-registered patients is much greater. The main cause is administrative delay (in extreme cases, up to sixty years!) in removing names from the register (Cobb and Miles, 1983). Inflation works in the individual doctor's interest because doctors are paid according to their list size.

Remuneration of general practitioners

The complex system of remuneration for general practitioners (DHSS, 1983e) has several geographical implications. The main element is a capitation fee which varies according to list size, weighted according to the age distribution of patients on the list. Patients over 65 and those over 75 entitle a doctor to higher capitation fees. Payment according to list size means that different practices in the same area are to some extent competing for patients.

In addition a basic practice allowance is paid to cover office equipment, telephone expenses and so on. Fees are paid for work out of hours, for items of service such as vaccination, cervical cytology tests, contraceptive and maternity services and for treatment to patients not on the list. To encourage doctors to join together in larger group practices there is a group practice allowance (£1,055 in 1984). Practice in under-doctored 'designated' areas is encouraged by allowances of £1,880 to £2,865 (in 1984). In addition, payment is made to rural practices for each patient on the list who lives more than three miles from the surgery. In sparsely populated rural areas

Table 3.1: The distribution of general practice sizes in England

Practice size	1961	1977
1 doctor	5,337	3,419
2 doctors	6,384	4,198
3 doctors	4,008	4,917
4 doctors	1,984	3,872
5 doctors	715	2,420
6 or more doctors	450	1,970

Source: National Association of Health Authorities, 1983, Section 3.5, p. 2.

the rural practice payment may contribute more than 10 per cent of income for some doctors (Lumb, 1983). This helps to counterbalance reduced incomes through smaller list sizes.

Group practices

Perhaps the most significant change in general medical practice in recent years has been the growth of partnerships of three or more doctors into 'group practices'. Table 3.1 indicates the scale of the trend. Group practice promises real advantages for most doctors: the professional support of close contact with medical colleagues, readily available cover for days and nights off duty, holidays and periods of illness, and the sharing of premises and other practice expenses, to say nothing of the extra financial allowance. Another related change has been a gradual acceptance of health centres: multipurpose premises housing the surgery of a group practice together with other primary health facilities.

Except in the most isolated parts of the country, single-handed or two-doctor partnerships are now the exception rather than the rule. In Norfolk, for example, 81 per cent of practices in 1951 consisted of one or two doctors, whereas by 1981 the proportion had dropped to 39 per cent. The single village doctor is now something of a rarity as general practitioners join forces to cover a wider area. Surgeries in outlying villages have been closed, to be replaced by a single surgery serving what was formerly the territory of several practices. In Norfolk there are at present about 400 parishes (containing about 170,000 people) with no doctor's surgery (Fearn, 1983). In West Glamorgan, Phillips' maps (Figure 3.2) clearly

Figure 3.2: The changing organisation and location of general
practitioner services, West Glamorgan, 1960–77

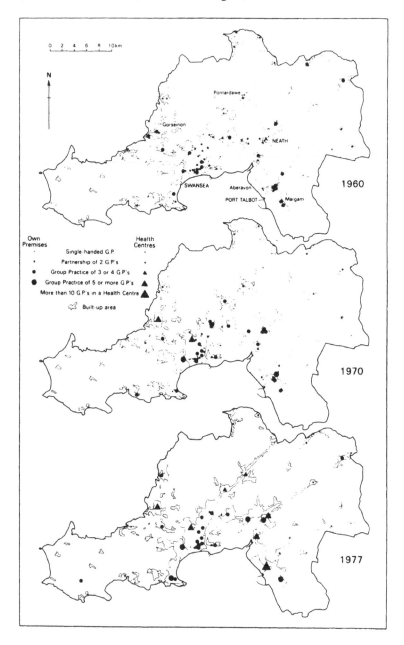

Source: Phillips, 1981, p. 90.

Table 3.2: Type of practice attended in rural and non-rural areas of the UK (percentages of rural and non-rural residents)

Informants attending	England	Wales	Scotland	N. Ireland	UK
Single-handed practice:					
Rural	8	30	20	12	11
Non-rural	20	12	11	28	19
4 or more doctors in practice:					
Rural	51	22	31	19	45
Non-rural	37	49	42	18	38
Practice with average list ≤ 1,800:					
Rural	13	25	30	10	15
Non-rural	9	8	22	20	10
Average list size over 2,500:					
Rural	36	33	11	36	33
Non-rural	49	22	19	25	45
Practice in a health centre:					
Rural	22	2	26	59	23
Non-rural	16	29	21	48	17
Base:					
Rural	776	60	113	59	1,008
Non-rural	2,800	143	278	60	3,281

Source: Ritchie *et al.*, 1981, p. 16.

illustrate a steady increase in the size of partnerships matched by a corresponding decrease in the number of surgeries (93 in 1960 falling to 49 in 1977) and growth in the number of health centres.

Table 3.2 compares the characteristics of practices attended by a national sample of 4,289 people. Practices are divided into rural and non-rural categories. 'Rural' in this context means electoral wards with a population density less than 15 per hectare which were also considered by interviewers to be rural in character. The table shows that far more people are registered with a general practitioner who belongs to a group of four or more doctors than with a family doctor practising single-handed, except in the rural parts of Scotland and Wales. Single-handed practices, perhaps surprisingly, are associated more with urban than with sparsely populated rural areas in England and Northern Ireland. They are most common, in fact, in poorer inner-city areas. The rural areas of England are dominated by group practices of four or more doctors, which has left many of the smaller settlements without a surgery.

Large practices in Scotland and Wales, by contrast, are more common in non-rural wards. Northern Ireland has not shared the growth in large practices to the same extent.

Patients registered with doctors who have small lists are most numerous in Scotland and Wales, particularly in rural areas. Large lists are most common in England (particularly in English urban areas) and least in Scotland. Health centres are used by about 20 per cent of all respondents in England, Wales and Scotland but by over half the respondents in Northern Ireland, where health centre policy has been vigorously promoted. With the exception of Wales, respondents in rural areas are more likely to use a health centre than those in non-rural areas. Health centres have tended to concentrate facilities that previously were more dispersed, so their popularity in rural areas has probably contributed to the loss of some services in outlying areas.

The result of the trend towards larger group practices and the closure of outlying surgeries has been an increase in the average distance people must travel to their doctor's surgery. Cartwright and Anderson (1981), reporting on the results of their national surveys of patients, revealed that the proportion of patients who estimated that they could get to the surgery in less than five minutes fell from 29 per cent in 1964 to 23 per cent in 1977. This deterioration in access time was in spite of the increase in patients using private transport to get to the surgery from 23 per cent in 1964 to 42 per cent in 1977.

Centralisation in surgeries

Two other trends have contributed to a substantial concentration of general practitioner services. Branch surgeries have been closing and home visiting has declined. Like the growth in group practices, both have affected the physical accessibility of primary care.

Branch surgeries are generally simple facilities located some distance away from the main practice premises and attended only a few times a week by the doctor. Their purpose is to extend the accessibility of general practitioner services to people who would find it difficult to reach the main surgery, some of whom might require a home visit if there were no local surgery. Branch surgeries exist in both urban and rural

Table 3.3: Distance of doctor's surgery from home, by whether practice has branch surgeries, in rural and non-rural areas (%)

Distance	Rural		Non-rural		All areas		
	With branch surgery	Main surgery only	With branch surgery	Main surgery only	With branch surgery	Main surgery only	All surgeries
< 1 mile	41	26	57	52	52	47	49
1–2 miles	18	18	26	30	24	28	26
2–5 miles	31	41	14	15	20	20	20
> 5 miles	10	15	2	2	5	5	5
Base	500	508	1,008	2,268	1,508	2,776	4,289

Source: Ritchie et al., 1981, p. 21.

settings, but their role in extending access to primary care is most critical in the more sparsely populated areas where distances to main surgeries are likely to inhibit trips by the less mobile groups in the population. Table 3.3 shows that rural residents are much more likely to have a long journey to the surgery than urban populations. In non-rural areas, the presence of branch surgeries has little effect on overall accessibility levels, but the effect is marked in rural districts. The table demonstrates that in rural areas the availability of a branch surgery increases the proportion of people less than one mile away from the surgery and decreases the proportion more than two miles away. As more branch surgeries close down, these benefits are lost. In Norfolk, for example, 19 new branch surgeries have been established and 53 have closed since 1951. New branches have been mainly in rural settlements with rapidly growing populations, often near to the towns. Closures have tended to occur in villages in the remoter parts of the county where the population has declined or has grown only slightly (Figure 3.3).

Although patients have more difficulty in getting to doctors' surgeries from rural areas than within urban agglomerations, home visiting by doctors is no more prevalent in rural areas than elsewhere (Ritchie et al., 1981). Home visiting is decreasing rapidly in all areas. There appears to be a danger that British general practice will become completely surgery-dominated, as it is in North America, that doctors will lose contact with the home environments of their patients and that all but the most immobile of patients will have to make the

Figure 3.3: Branch surgeries in Norfolk, 1951–81

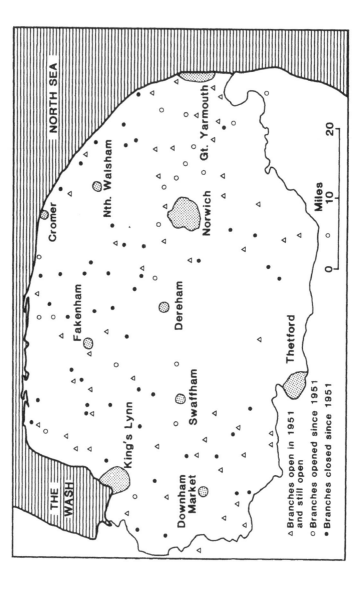

△ Branches open in 1951 and still open
○ Branches opened since 1951
● Branches closed since 1951

Source: Fearn, 1983, pp. 106–7.

effort to travel to the doctor. A study of home visiting by general practitioners in the north of England in 1969 was repeated in 1980 (Whewell *et al.*, 1983). In that eleven-year period the number of surgery consultations fell from 28.5 to 21.9 per doctor per day (a 23 per cent reduction). Home visits fell more dramatically, from 9.1 to 5.4 per doctor per day (a 41 per cent reduction). The decline was most pronounced in repeat visits, particularly to children, and least in visits to elderly patients. General practitioners still considered that about the same percentage of patients could have attended the surgery in 1980 as in 1969 (27 per cent of visits were judged 'unnecessary' in 1969 and 25 per cent in 1980). Cartwright and Anderson (1981), in a national survey, found that the proportion of consultations that were home visits fell from 22 per cent to 13 per cent in the period 1964 to 1977. They reported that people who normally used private transport to get to the surgery experienced the same decline in home visiting over the period, so the trend does not seem to be due to the increase in private transport. Of all the doctors questioned by Cartwright and Anderson in 1977, 52 per cent wanted to see less emphasis on home visiting in the future and only 6 per cent wanted more emphasis, so the trend is likely to continue. In spite of this, doctors who enjoy their work were found to be more likely to visit their patients than other doctors, and patients who received home visits reported more satisfaction with their doctor than other patients.

The location of surgeries

The most detailed description of the location of surgeries in British cities is that by Knox (1978, 1979), who examined Aberdeen, Dundee, Edinburgh and Glasgow. In Aberdeen he found a concentration of one-third of all the city's surgeries in a half square mile area on the fringe of the central business district. While one large area of older local authority estates contained several surgeries and the central middle-class area of the city was also well served, most medical manpower was localised in the mixed inner-city area. Most of the suburbs were badly served in comparison, with large tracts of both public and private housing having no local surgery at all.

Comparing Aberdeen with Dundee, Edinburgh and Glasgow,

Figure 3.4: Relative levels of accessibility to general practitioners' surgeries in Scottish cities

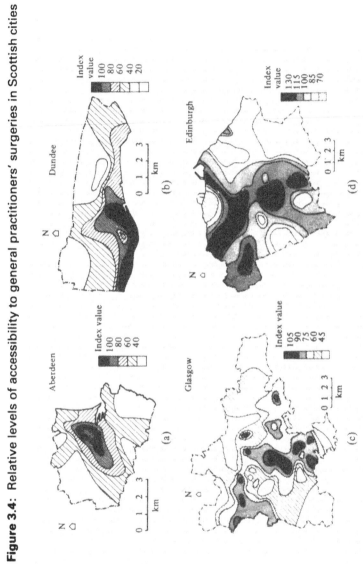

Source: Knox, 1978, p. 427.

Knox concluded that the location of surgeries tends to follow the distribution of shopping and business facilities outside the central business district itself. Surgeries do not avoid the deprived inner-city areas: indeed, they are particularly concentrated there. It is the post-war housing estates, both private and local authority, on the periphery of the cities that are least well served by general practitioner surgeries.

Knox pursued his analysis by calculating an index of accessibility to all surgeries in each city. The index was based on the size of surgeries and their distances from the place in question, weighted by expected modes of transport. Maps of the index (Figure 3.4) led Knox to suggest that the working-class areas on the periphery generally had the lowest levels of accessibility to primary care.

The present pattern of surgeries is largely due to locational inertia. Surgeries tend to be located in older residential districts, where parts of large houses have been converted into practice premises. Such districts have lost population in recent decades, but the post-war housing developments do not contain properties which are so readily converted and rarely have purpose-built surgeries. Middle-class areas remain attractive to doctors wishing to supplement their income by private practice, and because that is where the doctors themselves live. Some surgeries are held in the house of the doctor (one-third in Aberdeen in 1973), but this arrangement seems to be becoming less common as time goes on.

The concentrated nature of surgery locations noted by Knox may be against the interest of both patients and doctors. Clustered surgeries are inefficient in that they necessitate more travel by patients and doctors than a more even distribution. They also tend to produce large disparities in list size, with a few doctors in peripheral locations collecting excessive lists.

Locational preferences

Some areas in Britain have consistently found it difficult to recruit new doctors. The prospect of high workloads in some localities must act as a deterrent, especially where supporting services are weak. Conversely, some general practitioners may still be attracted to wealthier areas because of the opportunities there of supplementing their incomes through private practice. The presence of other doctors is a magnet because of the necessity to arrange cover in periods of absence. Doctors try to practise in areas where they already have interests and ties, including the areas in which they were trained. The regions with no medical schools or with few undergraduate places have more difficulty recruiting than other regions. Rural or coastal locations are seen as favourable characteristics. The attractiveness of areas is also related to the availability and condition of practice premises, suitable housing for the doctor's family and educational facilities for his or her family, as well as the social, cultural and professional climate of the district (Butler and Knight, 1975a). Low-income areas are avoided as practice there is seen as a real disadvantage for future career prospects (Hart, 1971).

Practitioners base their locational decisions not on accurate and objective information but on imperfect knowledge and personal values. Understanding of the present distribution and policies designed to guide future distributions must therefore be based as much on doctors' perceptions as on the objective characteristics of places. Pursuing such an argument, Knox and Pacione (1980) conducted a questionnaire survey of final-year medical students in Dundee and Glasgow. Respondents were asked to assess the importance of a series of factors in choosing a location in which to practise as a family doctor. The professional factors mentioned most frequently were accessibility to specialised hospital facilities, the availability of other doctors to cover for time off and the availability of suitable surgery premises. Social factors cited equally strongly were access to leisure and recreation facilities, the opportunity to work in a rural area and proximity to relatives and friends. Few students mentioned the opportunity to treat private patients, the eligibility for financial incentives or the opportunity to work with a particularly 'needy' population as positive reasons for choosing a particular place.

When the students were asked to evaluate the attractiveness of 16 Scottish regions in terms of both living and working, a consensus view emerged. There was general agreement on the attractiveness of Edinburgh, the Borders and Argyll and the lack of attraction of the far north and the industrialised central lowlands (where most of the under-doctored areas are). If new entrants followed these expressed preferences, their distribution would exacerbate the existing maldistribution.

Repeating the same exercise for different areas within Glasgow and Dundee produced the same conclusion on a more detailed scale. The areas with the lowest scores of attractiveness for general practice were the areas with the worst indices of deprivation, overcrowding, unemployment, lack of housing amenities and old-age dependency. These were either inner-city districts or public housing estates around the periphery. The most attractive areas were the most affluent owner-occupier districts, together with the areas close to the two medical schools. The scores for areas as workplaces were consistently less extreme than the scores for residential areas, suggesting that doctors might be induced to work in some places where they would not be persuaded to live.

Medical Practices Committee

If doctors were free to establish themselves in general practice wherever they chose, their distribution over the country would undoubtedly be uneven, as it was before the foundation of the National Health Service. One of the chief reasons for the improvement in coverage in England and Wales is the direction imposed by a body known as the Medical Practices Committee.

Of the nine members of the Medical Practices Committee, seven are general medical practitioners. The committee's function is to control the distribution of medical practitioners providing general medical services within the NHS throughout England and Wales. It considers the need to fill each general practitioner vacancy as it occurs and considers all new applications for inclusion in the medical lists of family practitioner committees. The Medical Practices Committee has the power to refuse an application if it considers that the number of doctors already practising in an area is adequate.

Each family practitioner committee administrative area is divided into smaller practice areas (there are about 1,400 in England) and these are classified according to the number of patients per doctor, after adding one more doctor. The classification is:

Designated areas with average list sizes of 2,500 and above
Open areas with average list sizes of 2,101–2,499

Intermediate areas with average list sizes of 1,701–2,100
Restricted areas with average list sizes of 1,700 or below.

Adjustments are made to exclude the practices of certain part-time and elderly doctors, and the committee may also consider special local circumstances, such as substantial new housing construction. Applications to practise in designated and open areas are normally granted without question and an initial practice allowance is payable for those setting up practice in designated areas. Doctors who practise in designated areas over a period of time also receive additional remuneration in the form of a designated area allowance. Applications for intermediate and restricted areas are considered on the detailed advice of the family practitioner committee and may be refused.

In Scotland the equivalent body is the Scottish Medical Practices Committee, which has six members of whom four are general practitioners. The Scottish system of negative direction is more flexible and liberal. Scottish medical schools produce proportionately more doctors than those of England and Wales, and list sizes in Scotland are correspondingly lower.

Areas in Scotland are classified only as designated or not designated according to whether the adjusted average list size is more or less than 2,500. This determines the payment of the designated area allowance. Adjustments are made by weighting rural patients more than non-rural patients (by taking account of rural practice payments) and by discounting doctors' commitments outside general practice. Initial practice allowances are made at the discretion of the committee whether the area is designated or not. Each application to practise is considered individually after consultation with the health board concerned. Applications may be refused if the list or average list of patients in the partnership would be below 1,500 after admission of the applicant. For sparsely populated rural areas, consideration is given to whether it is essential to maintain a practice.

Redistribution after 1948

The most significant power of the Medical Practices Committee is to prevent entry into the most desirable areas. This negative form of direction has worked best when the supply of doctors exceeds the death and retirement rate. In the early years of the NHS a large number of newly qualified doctors were entering the market. Many of them had no choice but to set up practice in under-doctored working-class areas. During the 1950s, the distribution of general practice was transformed as vacancies in the less attractive areas were gradually filled. In the early 1960s, however, the supply of general practitioners was reduced by medical schools decreasing their intake of students, the expansion of more prestigious medical posts in hospitals and the 'brain drain' to more lucrative jobs overseas, especially in North America. The relative shortage of general practitioners hit the least attractive areas hardest. From 1961 to 1967 the percentage of people in England and Wales living in designated areas doubled from 17 to 35 (Knox, 1979b).

Changes in general practitioners' remuneration were introduced in 1966. These discouraged the collection of large lists of patients by reducing the relative importance of capitation fees, encouraged the improvement of practice premises by offering subsidies and rewarded movement into designated areas by raising the allowances payable. In the late 1960s, medical schools again expanded their output of graduates and from this time the number of designated areas declined once more. Many of the vacancies in designated areas were filled by doctors from overseas. In 1969, for example, 97 per cent of new general practitioners entering designated areas were from overseas (DHSS, 1970a).

By 1972–3, 17 per cent of practice areas in England were designated. The regional distribution was uneven, with most designated areas in the conurbations of the north and Midlands and particularly in the North West standard Region. The South West, East Anglia and the South East had very low proportions of designated areas and high proportions of intermediate and restricted areas (Butler and Knight, 1976). Since then, the number of designated areas has gradually reduced. By mid-1985, only one practice area in England and Wales was classified as designated: Spennymoor in County Durham (Medical Practices Committee, 1985b). Judged by this crude criterion, the redistribution policy has been successful.

Evaluation of Medical Practices Committee policy

Butler and Knight (1975, 1976) have reviewed the problems of the redistribution policy, which has remained unchanged in principle since 1948. They observed that the number of doctors working in each medical practice area varied considerably, reflecting differences in population density and area size. While the majority of areas contained between 2 and 30 general practitioner principals, nine areas in the country contained only one doctor, and one area contained over 60 doctors. Paradoxically, the areas with fewest doctors were less likely to be designated and more likely to be restricted than areas with more doctors. This is partly because the areas with few doctors tend to be the more attractive rural areas with sparse populations where rare vacancies are soon filled. Additionally, in less attractive places, the rule of adding an additional doctor before calculating average list size makes it very difficult for areas with few doctors to be designated. An area with only one doctor would need an average list of at least 5,000 to qualify for designated status, for example, and an area with two doctors would need an average list of 3,750. One implication is that rural areas with sparse populations are unlikely to be designated, even if the average workload happens to be unusually heavy. A second is that a comparatively large number of doctors (20 on average) benefit from financial inducements in any one designated area.

Within medical practice areas there is considerable variation in individual list sizes. Designated areas frequently contain a substantial proportion of doctors with individual lists less than 2,500 who are none the less eligible to benefit financially from the status of their area. A national survey conducted by the Office of Population Censuses and Surveys found that 22 per cent of respondents in designated areas were registered with a general practitioner whose list (or average list if he was in group practice) was less than 2,500 (Table 3.4). On the other hand, 37 per cent of respondents in non-designated areas were registered on a list exceeding 2,500. Use of the average list size in a medical practice area is therefore a somewhat blunt instrument of policy.

In Butler and Knight's view, many medical practice areas are simply too large, sometimes covering a whole city. In large non-designated areas there may be substantial under-doctored

Table 3.4: Size of practices attended in designated areas and non-designated areas (%)

List size	Designated area	Non-designated area	All areas
< 1,800	4	12	11
1,801–2,100	3	17	15
2,101–2,500	15	29	27
2,501–3,000	37	24	25
> 3,000	40	13	17
Not known	2	5	5
Base	610	3,679	4,289

Source: Ritchie et al., 1981, p. 14.

parts, and large designated areas may contain neighbourhoods that are attractive to doctors. Since the boundaries of practice areas do not correspond with other health and social services areas, routinely collected data on population characteristics, health needs and existing services cannot be used to compare areas. The reason the boundaries are inflexible and revision of areas by the Medical Practices Committee has been difficult is that the financial allowances are tied to the existing areas, so any revision would have widespread implications for the remuneration of general practitioners.

Butler and Knight found that the most important financial incentive was the designated area allowance. The initial practice allowance, payable to doctors commencing practice in designated areas, was much less commonly paid. Most general practitioners regarded the designated area allowance as far too low to affect their decision in choosing a practice location. Following a reduction in average list size to below the threshold brought about by new practitioners entering the area, the allowance is lost after a concessionary period. This may be counterproductive, since new entrants to undesirable areas may lose the allowance after a time, and established doctors already in designated areas have an incentive to keep newcomers out.

The policy of negative direction (of refusing entry into restricted and intermediate areas) also has its problems. A review of the decisions of the Medical Practices Committee showed that it had not been strictly enforced. Other considerations (which are confidential) may override an area's restricted status. In 1973, for example, Butler and Knight (1975b) noted that as many as 45 per cent of new general practitioner positions went into 'restricted' areas. At its best, negative direction is hampered by the Medical Practices Committee's inability to amend boundaries and is a weak method of control when the number of general practitioners is declining.

Future policy options for redistribution

How might the redistribution policy be improved? Butler and Knight recommended that the designated area allowance should be withdrawn and replaced by a higher payment for a much smaller number of areas with a history of very large lists (over 3,000 patients). They suggested that new practice areas should be defined with about 20–30 general practitioners each and populations in the 50,000–70,000 range, with boundaries reflecting catchment areas and following health district boundaries as much as possible. The list size criterion should be examined critically and other indices of dispersion ought to be considered. They also suggested abandoning the idea of a national standard in favour of local autonomy in determining the geographical distribution of manpower needs. Since the conditions of practice are influential for many doctors, future initiatives should perhaps concentrate on improving the pro-fessional facilities in under-doctored areas rather than on providing increased allowances. Developing postgraduate medical centres and making clinical assistantships and general practitioner beds available in hospitals were seen as the most promising lines of attack for the less attractive areas.

A DHSS working party was also critical of the designated area allowances and recommended that they should be phased out (DHSS Working Party on Under-Doctored Areas, 1979). It observed that in extreme circumstances the allowances may be counterproductive as they give existing doctors a financial motive for resisting newcomers. For the working party, an 'under-doctored' area is where doctors are unable to work effectively because of constant pressure or where the range, quality and accessibility of services are unsatisfactory from the patient's point of view. Average list size was seen as an important contributor but other factors were thought to affect workload and the standard of medical care. There are those concerned with the morbidity of the population, such as its age, sex and ethnic composition and the physical and social characteristics of the local environment. There are also factors contributing to the effectiveness of the services provided, like the characteristics of doctors and their practice premises, the availability of other primary health services, social services and secondary health services.

After considering how all these factors might be taken into account in identifying under-doctored areas, the working party rejected the idea of using a statistical formula or index because of the difficulties in obtaining and updating information on a consistent basis and of weighting the various factors. Instead it recommended that they should all be considered 'in a less structured way' by family practitioner committees to recognise areas which have a heavy average workload for doctors. The working party suggested the workloads in areas covering 25 per cent of the nation's population might be regarded as 'heavy', and suggested that doctors in such areas should be eligible for an extended initial practice allowance scheme. This, it was thought, would encourage new entrants to take on small lists where necessary and would encourage more existing doctors to form partnerships and admit more partners.

These recommendations conform to the pattern of other reports written by doctors in that they attempt to improve the quality and quantity of medical services by offering general practitioners more remuneration. It is also noteworthy that the medical profession itself is the regulator of the scheme, through the Medical Practices Committee and the family practitioner committees, which themselves are enjoined to take into account the views of local practitioners. They also fail to grasp the nettle of how to take account of several factors simultaneously. To do this 'in a less structured way' than in a formula implies that the process of making decisions and assumptions about information requirements and factor weightings is done intuitively or unconsciously. How the results of family practitioner committee deliberations based on unrevealed and perhaps even unidentified criteria might then be equated at national level remains an unanswered question.

COMMUNITY NURSES

Domiciliary health care complementary to family doctor services is provided by district nurses, health visitors and community midwives. The general practitioner acts as co-ordinator of these services at a local level: he or she is the link in the primary care team. District nurses give nursing care to people, especially the elderly, in their own homes, and some-times give treatment in the doctor's surgery. Before 1974 district nurses were employed by the local authorities, but now they work for health authorities. In addition, some general practitioners employ practice nurses to give treatment in the surgery.

Health visitors are registered nurses with additional training who are specialists in preventive medicine and health educa-tion. Among other duties, they pay regular visits to see mothers and children from ten days old until school age, to advise on infant and family health.

Community midwives are less involved with deliveries than formerly, unless they are attached to a general practitioner maternity unit. Since the Peel Report recommended in 1970 that all mothers should have their babies delivered in hospital, where facilities are available to cope with unexpected compli-cations, the number of births at home has declined to less than 2 per cent of all births. Community midwives continue to provide ante- and post-natal care to mothers at home and, indeed, have a statutory duty to attend mothers with new babies for the first ten days.

Nurses working in the community were the subject of a recent survey of 24 health districts in England and Wales by the Office of Population Censuses and Surveys (Dunnell and Dobbs, 1982). In the districts sampled, one-third of nurses working in the community were district nurses and one-fifth were health visitors. Less numerous groups were nursing auxiliaries (11 per cent), midwives (10 per cent) and school nurses (8 per cent). Only 6 per cent of the 5,401 nurses represented were employed by general practitioners (27 per cent of the practices in the sampled districts employed nurses). The remaining nurses were family planning, community psychiatric, geriatric, liaison and clinic nurses, all in small proportions.

Dunnell and Dobbs found that there were variations in the

Table 3.5: Community nurses: variations in staffing levels and time spent on patients in 25 health districts

District rates	Average	Maximum	Minimum
Number of nursing staff			
per 100,000 children aged under 5	362	525	194
per 100,000 children aged 5–15	52	182	11
per 100,000 adults aged 65 or more	392	608	213
Hours per week spent			
per 100 children aged under 5	3.15	4.33	2.51
per 100 children aged 5–15	0.89	1.78	0.22
per 100 adults aged 65 or more	4.40	6.19	3.32

Source: Dunnell and Dobbs, 1982, pp. 76–7.

proportions of different types of nurses between health districts. District nurses, for example, could comprise as little as 25 per cent or as much as 42 per cent of the workforce, but there were generally more district nurses in the districts with high proportions of elderly population. Staffing rates varied considerably between districts. Table 3.5 gives the number of community nurses employed for each of three groups of patients in an average district, the best provided district and the worst provided district. For every 100,000 children under five, the number of health visitors ranged from 194 to 525, with an average of 362. School nurse rates are given for children aged 5–15, and these vary considerably from 11 to 182. In relation to the elderly population, the numbers of district nurses, auxiliaries and geriatric nurses have an almost threefold variation between districts.

These differences in staffing levels are substantial and can be expected to be even larger over the country as a whole, but classifying the 24 districts on the basis of their geographical characteristics did not reveal any trends. The rural or inner-city districts, for example, were not consistently well or poorly provided. The apparently arbitrary differences between districts were narrowed when the amount of time spent by health visitors, school nurses and district, auxiliary and geriatric nurses with their respective patient groups was considered, but the eightfold difference in the school nursing service between the best and worst supplied districts is still striking.

As community nurses see most patients in the patient's own homes they spend more time travelling than most other health

Table 3.6: District nursing: where patients are first seen (%)

	1975	1976	1977	1978	1979[a]
Patients aged under 5 years					
Patient's home	16	14	12	12	11
Health centres and GP premises	80	83	85	85	86
Patients aged 5–16 years					
Patient's home	—	19	17	16	15
Health centres and GP premises	—	79	81	82	84
Patients aged 17–64 years					
Patient's home	—	28	27	26	25
Health centres and GP premises	—	71	71	72	74
Patients aged 65 years and over					
Patient's home	73	71	70	68	68
Health centres and GP premises	24	27	28	28	29

Note: a. Estimates.
Source: DHSS, 1981c, pp. 107–10.

service personnel. Dunnell and Dobbs arranged for community nurses to record their activities in diaries and found that district nurses and community midwives spent about a quarter of their time travelling, half with patients and a quarter on non-clinical activities (meetings, clerical and administrative work). Health visitors spent less time travelling (16 per cent) and with clients (35 per cent) and more on non-clinical work (46 per cent). Nurses in rural areas did not travel more than those in urban areas; their travelling was about average. Neither did the type of attachment affect the proportion of time spent in travelling.

In recent years, there has been an increase in the proportion of district nursing work in health centres and surgeries and a corresponding decrease in the proportion of home visits. Even over a five-year period, the trend is clear (Table 3.6). What is not known is whether the additional people being seen in surgeries are relatively active people who were previously seen at home or a new group of people not seen previously (such as patients discharged early from hospital). If it is the latter, the pattern of care is shifting away from the housebound.

Nurses: attachment versus patch

Community nurses in the past usually covered a particular geographical territory or 'patch'. More recently many health authorities have attached community nurses to general practices. In this alternative arrangement, the nurse is responsible for the patients on a practice list. The aim of bringing together general practitioners and community nurses in primary health care teams is to ensure as far as possible that their services are co-ordinated and directed to where they are needed. With a national average list size of about 2,000 and present establishment levels of one health visitor and district nurse for every 4,000 or 5,000 people, this means that attached district nurses and health visitors will work with at least two general practitioners.

Over four-fifths of the community nurses surveyed by Dunnell and Dobbs (1982) were attached to general practice. Of these, the majority of nurses worked entirely with the patients on the lists of general practitioners to whom they were attached, but some worked in a particular geographical subdivision of their general practitioner's catchment area and others had geographical patch responsibilities as well as their practice attachment. Attachment was most commonly to one practice, but to cover as many as four practices or more was not unusual. Most community nurses worked with 4–6 general practitioners. Of the fifth of nurses not attached in any way to general practice, most worked entirely on a geographical patch system. Districts in rural areas and inner-city areas had fewer nurses attached to practices than average. It was among the nurses not attached to general practice that the survey found the lowest levels of satisfaction with the opportunities to get to know general practitioners and to discuss patients with doctors.

In some urban areas, however, the attachment scheme has been reversed and health authorities have returned to a geographical patch organisation, particularly for health visitors. The main reasons are the considerable overlapping of general practitioner catchment areas, which create wasteful visiting patterns, and the possibility of missing people who are not registered with a general practitioner (London Health Planning Consortium Study Group, 1981). A health authority survey in Paddington illustrates the extent of overlap between practice areas in a densely populated urban district (Table

Table 3.7: Use of community nursing services and GP registration among the residents of ten tower blocks in Paddington, London

Tower	Total number of households	Number of households known to community nursing staff (col. 2)	Number of GPs with whom households in col. 2 were registered
A	222	71	14
B	55	25	16
C	125	16	8
D	167	35	21
E	143	6	4
F	200	8	5
G	163	8	6
H	125	28	18
I	120	45	17

Source: London Health Planning Consortium Study Group, 1981, p. 56.

3.7). Of the ten blocks of flats investigated, the extreme example was a block in which community nurses visited 35 households registered with 21 different general practitioners. A situation in which the same street or even building is visited by several community nurses is clearly not using scarce resources in the most efficient manner. While this one *ad hoc* survey drew attention to a local example, it is not known how much overlap of practice areas generally occurs in inner cities, in suburban areas, in smaller cities and towns and even in rural areas. Still less is understood about the additional travelling effort for community nurses which attachment implies in all these types of area.

Not being registered with a general practitioner is most common in inner-city areas. People who are unaware of the system of registering with a family doctor, those who are unwilling to register and those who try but are unable to register tend to be socially deprived people whose need for health care is higher than average. The national rate of non-registration has been estimated at 2 per cent but could be as high as 30 per cent in parts of central London (London Health Planning Consortium Study Group, 1981). Community nurses who work solely from general practitioner lists would therefore miss a small but important group of potential patients. Some elderly people would not become known to the district nursing services until their condition became an emergency. Equally disturbing is the possibility that babies and young children at risk might slip through the network of health visitors. These risks are less when community nurses are responsible for a geographical patch.

HEALTH CENTRES

To facilitate co-operation and teamwork between general practitioners, district nurses and health visitors, the development of health centres has been encouraged. Health centres are premises provided by health authorities to be the focus of family health services in the locality. They contain surgeries rented by one or perhaps two practices of family doctors, consulting and administrative facilities for the home nurses and health visitors attached to the practice and perhaps also some specialised clinics and other health services.

The additional services could include ante-natal and post-natal clinics, pre-school and school health clinics, immunisation and vaccination, health education, family planning, speech therapy, chiropody, assessment of hearing, physiotherapy and community dental services. Independent contractors other than general practitioners may also be invited to rent part of the premises for general dental or ophthalmic practice and pharmaceutical business. Social services field staff might be offered interviewing and office facilities for their work connected with health care. Health centres can also be the location for selected out-patient clinics. Although out-patient clinics held outside the hospital use consultants' time in travelling and take consultants away from the supporting facilities of the hospital, they are acknowledged to foster the growth of mutual trust and confidence between general practitioners and consultants. From the patients' point of view, there are real advantages in seeing a consultant in familiar surroundings with the family doctor close at hand, and with a journey that is probably easier than that to the hospital. Clearly, the wider the range of services provided, the better opportunities there are for integration and promoting familiarity and use among patients.

Health centres have been seen by the DHSS as an appropriate way of upgrading primary health care in deprived areas where needs are high but facilities are below average. Inner-city areas have the highest priority for health centres, although it is acknowledged that deprived rural areas with sparse population and little public transport may be considered for small centres (DHSS 1979b). The size of centres is gauged by the number of general practitioners who have their surgeries there: normally between three and twelve, with an average around five. In rural areas, health centres may contain less than three doctors if the health authority is satisfied that the grouping of three or more doctors would mean that patients had to travel an unacceptable distance. According to the guidelines, the upper size limit is set specifically to avoid an institutional atmosphere and long travelling distances for patients. Health authorities are advised to consult community health councils to ensure that the most important factor to be taken into account when deciding the location of a health centre is accessibility to the public (DHSS, 1979b). No guidelines are given, however, on what travelling distances might be considered unacceptable, or how to measure accessibility.

The Department's enthusiasm for health centres, especially as a policy instrument for deprived inner-city areas, was backed by the earmarking of capital sums from 1974 for health authorities to use for construction and conversion. Financial assistance was also made available to general practitioners who wished to extend their own practice premises to accommodate other members of the primary health care team. The emphasis on spending was short-lived, as later the allocation of capital funds for this purpose was stopped and authorities were warned to consider carefully before embarking on a health centre building and spending spree (DHSS, 1980a).

Assessment

Following a spurt of building in the few years after 1974 when finance was made available, the establishment of health centres has slowed down in England and Wales. In Northern Ireland and Scotland, however, the policy has been pursued more actively. Scotland is notable for the scale of a few centres. There are two, in Woodside (Glasgow) and Clydebank, which each involve up to 30 general practitioners and cover populations as large as 70,000 (Levitt and Wall, 1984).

Health centres do not appear to be an attractive proposition to a large number of general medical practitioners who value their own professional and financial independence and are wary of putting themselves in a position of dependency in health-authority-owned premises. Furthermore, general practitioners on the whole like to have their own premises (with mortgages met by family practitioner committee funds) because they are a useful capital investment for retirement. Without a change in these attitudes, it is difficult to envisage another health centre boom.

What of the health centres which have already been established? According to the objectives of the policy, general practitioners in health centres should be working more closely together, with hospital doctors and with auxiliary staff, and should be able to deliver a higher quality of service to patients than practitioners working from conventional surgery premises. These propositions were investigated by Cartwright and Anderson (1981) in a national study of 365 doctors, 22 per cent of whom worked in a health centre. Their most striking finding was the relative lack of difference between doctors working in health centres and elsewhere. The average partnership size was similar in the two groups. Health centres in general had more facilities and equipment than conventional surgeries, but not markedly so. Formal links with hospitals appeared to be slightly less amongst health centre practitioners than the others, and there was no evidence of closer working relationships with hospital doctors. Although practices in health centres were more likely to have district nurses, health visitors, midwives and social workers attached than practices elsewhere, the differences were not large. Communications with the social services were described as 'good' by only one-quarter of general practitioners in health centres. Doctors in health centres did not rely any less on deputising services than other doctors, nor were they on call for more or less nights a week. Job satisfaction and attitudes towards patients were remarkably similar in the two groups.

Cartwright and Anderson also questioned 836 patients, to see whether any difference in satisfaction existed between those who attended conventional surgeries and those who used health centres. There was no evidence from this survey that patients felt any benefit from health centre organisation. Their replies suggested that, if anything, doctor—patient relationships were rather less good in health centres than in conventional surgeries.

DENTAL, OPHTHALMIC AND PHARMACEUTICAL SERVICES

While some treatment of teeth and eye disorders takes place in hospital and hospitals have their own dispensary departments for pharmaceutical supplies, most dentists, ophthalmic practitioners and pharmacists are independent professionals who make their services available to the public directly. Like general medical practitioners, they work under contract with the National Health Service through their local family practitioner committee (or health board in Scotland), which provides them with a fee for each item of service. While family practitioner committees and health boards can influence the distribution of medical practitioners in their advice to the Medical Practices Committee, they have no equivalent power over general dental contractors, pharmacists or suppliers of ophthalmic services. Members of these professions are free to set up in practice wherever they choose. Their contract does not guarantee a minimum living income (unlike general medical practitioners), so they are more likely than family doctors to practise in densely populated areas where they are accessible to large numbers of customers. Areas of low population density have little attraction for them.

Some dental and ophthalmic services in addition are provided as part of the school medical service for children. These practitioners work for the health authority. Screening of schoolchildren takes place irrespective of location, although any treatment which is found to be required is a different matter: it must be obtained from independent contractors in the usual way. Many dental and ophthalmic contractors undertake private work as well as NHS work, and most pharmacists combine their dispensing activities with retail sales in a chemist's shop.

General dental practitioners

Dentists do not have a list of patients. Strictly speaking, their responsibility for a patient ends when a course of treatment is completed. Although patients usually return to the same dentist for subsequent courses of treatment, there is no obligation on either side. Compared with the distances people travel to their doctor, distances to the dentist are high (Table 3.8).

Table 3.8: Distance travelled to the dentist, by rural and non-rural areas (%)

Distance	Rural areas	Non-rural areas	All areas
< 1 mile	22	43	38
1–2 miles	13	24	22
2–5 miles	25	21	22
> 5 miles	36	11	17
(with nearer dentist)	(21)	(10)	(13)
(no nearer dentist)	(15)	(1)	(4)
Base	318	1,064	1,382

Source: Ritchie *et al.*, 1981, p. 125.

Ritchie's survey found that journeys exceeding five miles were common in rural areas. In non-rural areas over a tenth of journeys were more than five miles. This is often through choice rather than necessity. Of the people travelling more than five miles, the majority in rural areas and the vast majority in non-rural areas said there was a dentist nearer than the one they visited. The reasons most frequently given for preferring a more distant dentist were that they were satisfied by the treatment he gave or that he had been recommended to them.

As dentists are not constrained by statutory restrictions, their choice of where to practise is entirely determined by their personal assessment of the opportunities available and their preferences for certain areas or types of place. Apart from the obvious effect of areas with large populations tending to attract dentists, several other factors have been identified as influencing the decision, including the affluence of the community (Cook and Walker, 1967) and the presence of family ties (Robinson *et al.*, 1980). One of the strongest influences is the location of the university in which the dentists trained, as is illustrated by Thexton and McGarrick's (1983) study of the practice locations of dentists graduating from English, Scottish and Welsh dental schools in the period 1975–80.

Thexton and McGarrick produced separate maps to show the location of dentists according to the universities they attended (the five dental schools in London were considered as one). These maps (Figure 3.5) show that there is a marked tendency for dentists to practise near their dental school. The

Figure 3.5: The distribution of dentists according to the dental school attended

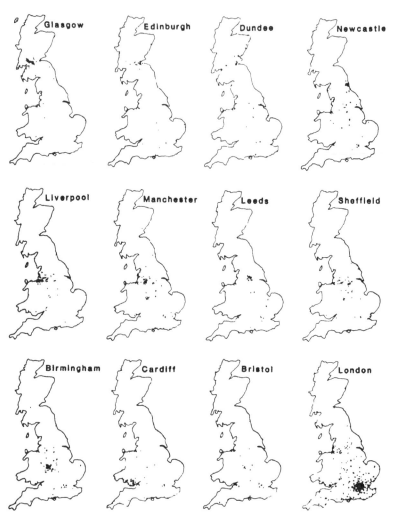

Source: Thexton and McGarrick, 1983, p. 72.

distribution of dentists around each dental school is described by a mathematical power function in which the rate of diminution with distance was found to vary between dental schools; a reflection perhaps of the varying distributions of opportunities around the schools. Only 18 per cent of Dundee graduates were practising within 40 kilometres of Dundee, while 70 per cent of Glasgow graduates were within 40 kilometres of central

Glasgow, to take the two extreme examples. In addition to the strong tendency for graduates to find work near their universities, Thexton and McGarrick also found that the northern universities (Dundee, Glasgow, Edinburgh and Newcastle) were net exporters of dental manpower to central and southern regions of the country.

Regional distribution of dentists

Dental services follow the 'inverse care' law discussed in Chapter 2. The general distribution is a concentration of dentists in southern England and more favourable dentist to population ratios in urban than in rural areas over the remainder of the country. There is a clear association with social class. Cook and Walker (1967) reported that areas with high proportions of persons in social class I had high numbers of dentists. The school dental service and private dental practice have similar geographical distributions and reinforce the uneven pattern of general dental services.

There are signs that the large imbalance in the distribution of dentists is improving gradually through time. While in 1974 there were 2.3 times as many dentists in the best provided region compared with the worst (after taking population size into account), the differential was reduced to 2.2 in 1977 and 2.1 in 1983. Table 3.9 gives the detailed figures. Health authorities have no control over these imbalances, but they are directly responsible for the distribution of dentists they employ as part of the community health service. Here, too, there are variations in the dental staff per 100,000 population, with a differential of 1.6 between the worst served English region (Trent) and the best (Wessex). Wales, in a reversal of the expected pattern, has a better supply of community dentists than Wessex (DHSS, 1983b).

Carmichael (1983) has drawn attention to the considerable variations in dental services within a region. Some of his evidence is given in Table 3.10. The lower part of the table contrasts the Northern Region with the South Western Region, which has a similar population size, and also with England and Wales as a whole. The low provision of general dental practitioners in the Northern Region is matched by low values in measures of dental treatment there compared with

Table 3.9: Population per general dental service practitioner by region, England and Wales

Regional health authority	Population per dentist		
	1974	1977	1983
NW Thames	2,567	2,516	2,141
SW Thames	2,925	2,797	2,517
South Western	3,543	3,354	2,998
SE Thames	3,667	3,446	2,978
NE Thames	3,739	3,602	2,901
Wessex	3,784	3,667	3,383
Oxford	4,195	3,957	3,337
Mersey	4,654	4,135	3,827
Yorkshire	5,023	4,584	3,807
East Anglia	5,135	4,471	3,858
North Western	5,174	4,742	3,853
West Midlands	5,415	4,779	4,138
Wales	5,453	4,792	4,000
Northern	5,725	5,456	4,566
Trent	5,916	5,345	4,595

Source: Scarrott, 1978, p. 360; Central Statistical Office, 1985a, p. 71.

Table 3.10: Dentists and dental treatment statistics (1978/80) in the Northern Region

Division	Persons per dentist	Number of estimates per 1,000 population	Fillings to extractions ratio	Deciduous teeth conservation courses age 5–14 per 1,000 population
Newcastle upon Tyne	3,418	1,653	4.06	341
North Tyneside	3,860	1,816	3.30	347
Cumbria	4,233	1,791	2.88	315
Northumberland	5,163	1,444	2.85	276
Gateshead	5,738	1,376	2.87	236
Cleveland	6,114	1,521	2.26	297
Durham	6,348	1,308	2.29	258
Sunderland	6,400	1,121	2.08	196
South Tyneside	7,391	1,015	1.84	159
Northern Region	5,187	1,468	2.62	276
England and Wales	3,837	1,776	4.49	434
South Western Region	3,225	2,155	4.00	545

Source: Carmichael, 1983, pp. 337–8.

England and Wales in general and the South Western Region as an example of a better provided region in particular. The number of courses of treatment ('estimates'), the ratio of fillings to extractions (a measure of conservative as opposed to last-resort treatment) and the ratio of conservation courses for children are given. These show the Northern Region to be operating at roughly 60–80 per cent of the national levels of treatment.

Differences between districts within the Northern Region appearing in the upper part of Table 3.10, are even greater. The south Tyneside district has more than twice the population per dentist than Newcastle upon Tyne. There is a clear relationship between the number of persons per dentist and the amount of treatment supplied, with treatment levels in South Tyneside about half those in Newcastle and North Tyneside. North and South Tyneside are adjacent districts divided by the River Tyne, as are Newcastle and Gateshead. Carmichael's evidence draws attention to the sharpness of contrast between services on either side of the river.

Carmichael has no doubt that low levels of dental treatment are the result of low supplies of dentists, and not that some areas have an inherently low demand. If more dentists could be attracted to districts like South Tyneside and Sunderland, the demand for dental treatment there would respond in proportion. There seems little hope that current imbalances will correct themselves without some sort of incentive scheme or restricted area policy.

Several policy suggestions have been made to the Department of Health and Social Security. The Dental Strategy Review Group report (1981) recommended that financial encouragement for dentists to practise in the areas of greatest need should be considered. Some form of initial practice allowance or direct reimbursement of expenses were suggested mechanisms. Secondly, although recommending an overall 10 per cent reduction in undergraduate admissions to dentistry, the report suggested that student numbers should be maintained in dental schools in regions with relatively few dentists. Thirdly, for areas not well served by general dental practitioners the report suggested that routine dental care should be provided by the community dental service, for adults as well as children.

Pharmacies

Most patients obtain their drugs or medicine by taking a prescription written by their doctor to a pharmacy, or 'chemist', where the items are dispensed. Patients may ask their doctor to supply drugs and medicines if they would have serious difficulty in obtaining them from a chemist because of distance or inadequate communications and (except in Scotland) if they live in a rural area more than one mile in a straight line from any chemist. Dispensing by general practitioners must be approved by the family practitioner committee or the health board and the central Rural Dispensing Committee (DHSS, 1983d). About 5 per cent of the population is registered with a dispensing doctor (Ritchie *et al.*, 1981).

The heaviest users of dispensing services, whether from pharmacists or general practitioners, are the same groups with the highest frequency of general practitioner consultations: the elderly, women and people in social classes III, IV and V. The number of prescriptions given is closely related to the number of consultations. Most consultations with a general practitioner result in written prescriptions but some repeat prescriptions are made without the doctor actually seeing the patient. For the majority of the population whose doctor does not dispense, a consultation with the doctor usually involves a trip to the chemist as well. Over 80 per cent of people questioned in the Office of Population Censuses and Surveys' primary care survey had received their last prescription in person at their doctor's surgery (Ritchie *et al.*, 1981). Patients were more likely to go usually to a pharmacy near the surgery (60 per cent) than near home (31 per cent) or not near either (9 per cent).

Pharmacies not only dispense medicines prescribed by general practitioners but also sell drugs, medicines and appliances over the counter without prescription. It has been estimated that self medication accounts for twice as many items of medicine consumed as prescribed medicine, but not necessarily twice in terms of cost (Dunnell and Cartwright, 1972). The chemist is also used as an independent source of advice about health problems. Ritchie *et al.* (1981) found that 15 per cent of their respondents said they had asked for advice in a chemist's shop instead of their doctor in the previous year. Most of the complaints on which advice was sought seemed to be of a minor nature.

Using 1976 figures given in Ritchie *et al.* (1981), the average number of people per pharmacy was 4,917 in the United Kingdom. There is approximately one chemist's shop for every two general practitioners. In England, there is a higher average population per pharmacy (5,170), with the Midlands being the region with the highest average population per pharmacy (6,176). The other, less densely populated parts of the UK have lower population thresholds: 4,475 in Scotland, 3,981 in Wales and 2,850 in Northern Ireland. Chemists' shops in Northern Ireland operate with less than half the population catchment of shops in the English Midlands.

Access to pharmacies

About 86 per cent of the people interviewed in England, Scotland and Wales lived within two miles of a chemist's shop. Surprisingly, the proportion was less (74 per cent) in Northern Ireland, where there are more pharmacies per head of population, but this is explained by the disparities in the distribution of population and chemists in urban and rural areas there. Nationally there is a marked difference in the distance to pharmacies between rural areas (those with a population density less than 15 per hectare) and non-rural areas (Table 3.11). While almost all the non-rural people interviewed lived within two miles of a chemist, only half the people in rural areas had a chemist within this distance. The proportion of people whose doctors dispensed medicine increased with distance from the pharmacy in rural areas. Nevertheless, the majority of people living more than five miles from the nearest pharmacy said that their doctor did not dispense. The general practitioner dispensing service fills some of the gaps in the cover of pharmacies in rural areas, but perhaps only a minority of the gaps.

Most people do not feel that living at some distance from the nearest pharmacy is a particular disadvantage. Only 21 per cent of those 2−5 miles from a chemist's shop said it was fairly or very difficult to get there. Of those more than five miles away, 36 per cent said it was difficult. The people who had difficulty in getting to the chemist were likely to be elderly,

Table 3.11: Distance of nearest pharmacy from home, by whether rural or non-rural area (%)

Distance	Rural		Non-rural		All areas	
< 1 mile	34		81		70	
1–2 miles	19	(11)	15	(1)	16	(4)
2–5 miles	30	(30)	2	(12)	9	(27)
> 5 miles	16	(44)	0		4	(45)
Base	614		1,960		2,574	

Note: Figures in parentheses represent the percentage of each group who said their doctor normally supplied drugs or medicines.
Source: Ritchie *et al.*, 1981, p. 89.

with restricted mobility and those relying on public transport. As with access to the general practitioner's surgery, it is the people who are likely to need the service most who tend to have the greatest difficulty reaching it. Some are able to avoid the journey. In Ritchie's sample, 11 per cent of people who had received a prescription in the last year had it taken to the chemist for them by someone else. For women over 75, the proportion getting help was 43 per cent. It seems likely that the number of people who obtain a prescription but are not able to get their medicine is very small, but no study appears to have made an estimate.

The accessibility of chemists' shops in rural areas has been gradually declining. The Royal Commission on the NHS (1979) noted that the number of pharmacies had fallen by more than one quarter between 1963 and 1979. From 1970 to 1980, the number of chemists' shops in England and Wales was reduced from 11,479 to 9,401, a loss of over 200 per year over the decade. The fall in numbers was accompanied by a decrease in the proportion of chemists' shops with a low prescription turnover (DHSS, 1982; Welsh Office, 1983). Since 1980, however, the trend appears to have reversed, with a slight rise in the number of registered pharmacies every year up to 1985 (Pharmaceutical Society of Great Britain, 1985). Chemists have the problems of small shops for they, too, must compete with supermarkets for most of their non-dispensing trade. Their overheads are much higher than other small shops since they must employ a qualified dispensing pharmacist. To give some protection to pharmacies with a limited turnover which provide a valuable service in rural areas, an income supplement is available. To qualify, a pharmacy must dispense between 6,000 and 24,000 prescriptions a year and must be at least two kilometres from any other pharmacy. Payment is on a sliding scale according to the number of prescriptions dispensed, with a maximum of £1,200 in 1984 for the pharmacies operating closest to the margin (DHSS, 1984e). In spite of this support, closures continue. Evidence submitted to the Royal Commission by the Pharmaceutical Services Negotiating Committee included the estimate that pharmacies dispensing less than 23,400 prescriptions per year would not recover their costs, but would be kept in business only by their retail trade (which was said to be declining). There are no mobile pharmacies in Britain. Perhaps they are not considered commercially viable, but the most compelling reason for their absence is that they are unlawful. Under the Medicines Act 1968, all pharmacies must be in registered premises, which excludes vehicles (Moseley and Packman, 1983).

Pharmaceutical services in Scotland

In Scotland, the number of pharmacies has also been declining. Calder (1983) gives the number of chemists in Scotland as 1,724 in 1956, falling to 1,117 in 1981, in spite of an increase in the prescription rate from 4.2 to 6.6 per head per annum in the same period.

In sparsely populated areas of Scotland, about £70,000 financial aid per year is paid to pharmacies that dispense less than 1,800 prescriptions per month and are not less than two miles from the nearest pharmacy. At present about 100 pharmacies receive a share of this allowance. In areas too sparsely populated to support even a subsidised full-time pharmacy, medicines are obtained from a few part-time pharmacies, through collection and delivery services or from dispensing doctors. Collection and delivery arrangements consist of patients and pharmacists communicating through a mutually convenient collection point. Patients leave their prescriptions and later pick up their medicines from the collection point. Such arrangements may be officially recognised by the health boards, which then meet the transport costs incurred, but often they operate unofficially.

Dispensing by doctors is rare in Scotland. It accounts for

only 3 per cent of prescriptions and only about 130 doctors do it (Calder, 1983). The Scottish regulations state that a patient must normally get his or her prescription dispensed by a pharmacy. Only in the last resort can the health board require a doctor to dispense.

Ophthalmic services

The term 'optician' is often used for convenience to cover three professions: the ophthalmic medical practitioner, who tests sight, the dispensing optician, who supplies glasses, and the ophthalmic optician (the most common type), who does both.

Two-thirds of adults in the United Kingdom have spectacles prescribed for them sometime in their lives. It is considered advisable for people who do not wear spectacles to have a sight test every five to ten years, and more frequently with increasing age. People with spectacles are advised to have sight tests every two to five years, and more frequently after the age of 50. Ritchie and her colleagues found that only 17 per cent of the non-lens wearers questioned nationally had had a sight test in the previous five years. Of lens wearers, 82 per cent had been for a sight test within five years but the proportion was lower among the elderly. Most people who had not had a test recently did not feel one was necessary. Among the non-spectacle wearers only 1 per cent said that the inconvenience of the optician's location was the reason for not having a test, but 10 per cent of spectacle wearers who had not been for a recent test said that inconvenient location was the reason.

In England there is one ophthalmic specialist or optician for approximately every 6,000 population. Of these, over 60 per cent are ophthalmic opticians, about 30 per cent dispensing opticians and about 10 per cent are ophthalmic medical practitioners. There are therefore fewer opportunities for sight testing than for obtaining prescribed lenses.

The distances people travel to obtain ophthalmic services are greater than for any other type of primary health care. Table 3.12 shows that 17 per cent of people who had had a sight test in the previous five years had travelled more than five miles for it. In rural areas, 40 per cent had travelled more than five miles. This table omits those not having a sight test at

Table 3.12: Distance travelled from home or work for last sight test in rural and non-rural areas (%)

Distance	Rural areas	Non-rural areas	All areas
< 1 mile	19	41	36
1−2 miles	12	25	22
2−5 miles	25	23	23
5−10 miles	27	7	11
10−20 miles	9	2	4
> 20 miles	4	1	2
Not known	3	2	2
Base	368	1,243	1,611

Source: Ritchie *et al.*, 1981, p. 106.

all, so it may be biased with a larger proportion of short distances than would be representative of the population as a whole.

Although domiciliary visits for sight testing are available under the National Health Service, Ritchie's survey found that less than 1 per cent of the people questioned had ever had a sight test at home. Domiciliary sight tests are an extremely rare occurrence, even amongst the elderly or those with restricted mobility. Lack of knowledge of the service appears to be responsible.

OTHER PRIMARY HEALTH SERVICES

Other community (non-hospital) services are available as well as those of the primary care team, the school health service and community dentistry. Health authorities employ specialists in community medicine known as 'community physicians' whose role is to identify the health needs of the local population and suggest ways of meeting those needs. This usually involves supervising a statistical service, advising local authorities on environmental health and the control of infectious disease, promoting preventive medicine and health education, planning services and co-ordinating health services with the social services of local authorities.

Services such as speech therapy and chiropody are sometimes provided in hospitals, clinics or health centres, but NHS speech therapists and chiropodists also make home visits.

Speech therapists work in the community mostly with children whose speech is delayed. Chiropodists deal almost exclusively with elderly people. Chiropody is unusual among the health services because people seeking it are more likely to go to the private sector than the National Health Service. Ritchie and her colleagues (1981) found that only one-third of the patients in their sample who had had chiropody treatment received it from the NHS. The NHS chiropody services are clearly not able to meet the existing need. Private chiropody goes some way to fill the gap, but a social class gradient in the use of private chiropody treatment indicates that there is likely to be considerable unmet need among the manual group. Surveys of local opinion on services in rural areas have revealed widespread and serious concern over the difficulty of obtaining chiropody (Moseley and Packman, 1983).

Outside the National Health Service, there is a wide range of volunteer groups providing services related to health. At one extreme are the formally constituted bodies like Age Concern, the National Council on Alcoholism, the Spastics Society, MENCAP, MIND, the NSPCC and the Women's Royal Voluntary Service, together with the Red Cross and the St John's Ambulance Brigade. Many of them employ professional staff and some undertake agency work for statutory authorities. Then there are mutual aid groups which provide moral support or practical help for fellow sufferers. Examples are Alcoholics Anonymous, Combat (for sufferers of Huntington's Chorea), the National Schizophrenia Fellowship, stroke clubs, and so on. While these voluntary groups concentrate on particular targets, other care groups are organised on a neighbourhood basis, with the aim of promoting mutual help within a particular geographical area. Neighbourhood care groups are particularly valuable in monitoring the needs of the most isolated and vulnerable members of the community. Visiting the elderly is their main activity.

The most significant source of voluntary support for the health services is the least organised: the informal care provided particularly by families but to a lesser extent also by friends, neighbours, fellow workers, church members and so on. Informal carers bear a burden with high financial and social costs to themselves. Their services are critical to the feasibility of community-based health care. Without them it would be difficult if not impossible to avoid long-term institutional care for the dependent elderly, for mentally ill people and the mentally handicapped. Early discharge from hospital after treatment, day surgery and out-patient treatment are also made possible by informal care at home. Informal care has a geographical dimension in a mobile society where the generations are increasingly separated. Health services in retirement areas are under pressure not only to cope with higher than average proportions of elderly people but also partly to compensate for the remoteness of family support.

Access to primary care

Primary care under the National Health Service is not uniformly distributed over the country. Although real improvements in geographical coverage have been made over the long term, the recent trend is for key parts of the service to become more centralised. Family doctors have combined together in large group practices and health centres. Outlying surgeries have been closed and fewer home visits are being made. Pharmacies are also tending to close down in rural areas. The traditional independence of the family practitioner services has sometimes hindered the achievement of a more equitable locational pattern. The personal and professional objectives of doctors, dentists, opticians and pharmacists do not always coincide with the public interest, and the locational pattern of practice that is the most attractive to the profession is not the most equitable socially. Professional and administrative independence has not helped liaison with other community services, notably the nursing service. To be fully effective, primary health services must work together.

4

The Location of Hospitals

The geographical pattern of hospital services in Britain is changing. Although the distribution of population has shifted in recent decades away from urban areas to the suburbs and rural districts, the main impetus for change in the distribution of non-psychiatric hospital services has been in the opposite direction. Hospital location policy has been dominated by considerations of function, internal organisation and size. The tasks a hospital is expected to perform and the staff and physical facilities that are thought appropriate to meet those tasks have been the main determinants of hospital size. The size of a hospital in turn determines its catchment area and, indirectly, its location. An explanation of the evolving pattern of hospital locations must therefore refer back to the debates about the proper functions of hospitals and the best organisational means of providing them.

The historical legacy

Before the National Health Service was established there were two main groups of hospitals: municipal hospitals and independent voluntary hospitals. Municipal hospitals, financed from the local rates, had evolved from Poor Law infirmaries, which themselves had developed from parish workhouses. Voluntary hospitals depended on voluntary subscriptions, donations, endowments and payment by patients. Both groups contained enormous variations in size, function and medical sophistication, from small cottage hospitals providing accommodation for the chronically sick to large general hospitals catering for acute medical and surgical conditions (Abel-Smith, 1964).

When the National Health Service Act came into force in 1948, most of the voluntary hospitals in the country (including all the 'teaching' hospitals connected to university medical schools) and all the local authority hospitals were taken into national ownership. At the start there were 1,143 voluntary hospitals and 1,545 municipal hospitals. Over half the buildings were more than fifty years old, and most were ill-equipped, even by the standards of that time (Abel-Smith, 1978). The large general hospitals mostly occupied inner-city sites in districts which were to experience a substantial exodus of population. Psychiatric hospitals, by contrast, were often in rural settings in the buildings of former workhouses. The regional distribution of hospitals was far from even, as it reflected the pattern of private donations and legacies and municipal wealth. London was especially well supplied, as was Liverpool, where a local tradition of philanthropy and benefaction'had built up a large number of hospitals. The quirks of local history produced a distribution over the country with no plan and no system, with the result that often the best hospital facilities were available where they were least needed, as Aneurin Bevan reported to the House of Commons when he introduced the National Health Service Bill (Open University, 1977).

The early National Health Service policy for hospitals was to maintain and improve the hospitals which it had inherited. Resources were allocated in proportion to a region's existing commitment, which tended to perpetuate the differences in provision between regions. As late as 1960, the number of acute hospital beds per thousand population ranged from 3.0 in East Anglia to 5.6 in the Liverpool Region. Local differences were greater. Another source of concern was that the enormous variations in size, physical condition and function from hospital to hospital had also persisted. An overall strategy was badly needed.

POLICIES AND TRENDS

District general hospitals

The Hospital Plan of 1962 introduced by the Minister of Health, Enoch Powell, set national policies which were to guide hospital building programmes (Ministry of Health, 1962). A ratio of 3.3 beds per thousand population was to be the target level of provision, with an additional 0.58 beds for maternity, 1.4 beds for geriatric, 1.3 beds for mentally handicapped and 1.8 beds for mental illness patients for every thousand population. Hospital services were to be provided in large district general hospitals (of about 600 to 800 beds), each serving a population of 100,000 to 150,000. In addition, there would be some smaller units provided for maternity cases, long-stay patients, mentally handicapped and mentally ill patients. Most of the existing mental hospitals, especially the large ones and those in isolated positions, would eventually be closed down.

A few years later, the report of the Bonham-Carter Committee was yet stronger in support of the development of large hospitals (Central Health Services Council, 1969). Bonham-Carter's guiding principle was that each of the major in-patient specialties should be organised around a team of not less than two consultants (to ensure professional support and cover for off-duty periods). To justify employing two consultants in every major specialty, a catchment of 200,000 to 300,000 population was required, which suggested a size of 1,200 to 1,800 beds for a district general hospital. Most small hospitals would then be redundant, except that a number of them might be required to provide nursing care for patients already treated at the district general hospital and to be hospital 'outposts' in especially remote areas.

Other reports on particular hospital services endorsed the advantages of large hospitals. The Platt Report on accident and emergency services (DHSS, 1962) recommended that these should be centralised in specialised departments capable of providing a full 24-hour service and covering a catchment of about 150,000 people. The Peel Report on midwifery and maternity bed needs recommended that all confinements should be in hospital and that small isolated obstetric units run by general practitioners should be replaced by larger combined general practitioner and consultant units in district general hospitals (DHSS, 1970b). This was to ensure that ready access to consultant cover would be available on the grounds of safety for mother and child. There was some evidence from the research literature that the perinatal mortality rate (still-births and first-week deaths) was significantly higher where general practitioner obstetric units were used than in practices with no access to local maternity units (Hobbs and Acheson, 1966).

Another impetus came from the Lewin Report on the organisation and staffing of operating departments (DHSS, 1970c). This investigation found that most operating facilities were in small hospitals with one or two operating theatres, or else in larger hospitals with a number of theatres located in different departments. Such a dispersion of small-scale theatre facilities meant that the use of staff was very inflexible, work tended to be unevenly distributed and the use of some theatres was inefficient. One of the main difficulties was the provision for emergency surgery. Reliance on on-call and stand-by duties was thought to be imposing an intolerable burden, particularly in small hospitals with less than 100 beds and one or two theatres. It was a major source of dissatisfaction for theatre staff and was thought to be seriously hampering the recruitment of theatre nurses. The main recommendation was that operating services should be concentrated 'as far as possible', with no upper limit being mentioned. Emergency operations should be even more concentrated, into operating suites with sufficient demand to be devoted exclusively for emergency work.

The district general hospital policy has remained consistent in its main objective since 1962. The aim is eventually to provide a network of about 250 district general hospitals of about 600–900 beds. Each is intended to have specialised surgical and medical facilities and departments for maternity, psychiatric, geriatric and child patients. Some district general hospitals will have accident and emergency units, ear, nose and throat and ophthalmology wards in addition. The larger ones will also have the rarer regional specialties such as neurosurgery (DHSS, 1977).

Originally it was planned to start work on 90 new and 134 substantially remodelled hospitals in 1970–1, but serious underestimation of new hospital building costs caused the start to be more modest. Nevertheless, by 1974 about a quarter of district general hospital beds were provided in new or substantially remodelled hospitals (Abel-Smith, 1978). At about this time the national economy was ailing and it was becoming clear that the original building plans were too ambitious. The DHSS took up the concept of a 'nucleus hospital' of about 300 beds, which would be cheaper to build and run than a fully developed district general hospital but which would be capable of expansion later. Several nucleus hospitals were built in the late 1970s. At the same time, a number of new large district hospitals were finished, but could not be fully used because health authorities found they could not afford the running costs. A few remained empty for long periods.

Accident and emergency services policy

Of all the hospital services, the accident and emergency service depends most for its success on good physical accessibility. The policy governing the location of accident and emergency services is therefore of interest. It began with the Platt Report (DHSS, 1962). This report reviewed the 789 casualty departments then existing in England and Wales, most of them in acute hospitals. Even very small hospitals had casualty departments. Altogether, 38 per cent of casualty departments were in hospitals with less than 50 beds. Most departments operated on a small scale: 59 per cent dealt with less than 100 patients per week, of whom only about 4 would be seriously injured. Almost a third of them had no access to a fracture clinic. Consultant involvement was low, with most patients being seen only by junior hospital staff. There was widespread agreement that there were too many small departments struggling to provide an accident service often in inadequate accommodation and with shortages of medical, nursing and auxiliary staff.

One of the main problems, according to the Platt Report, was that casualty departments were almost overwhelmed by minor accident cases which could have been treated by a general practitioner. It found that of all patients treated in casualty departments, only 12 per cent had sufficiently serious injuries to cause admission as an in-patient or referral to an out-patient department or another hospital. These were the patients who should be given a better service. Platt's committee accordingly recommended that patients with serious and minor injuries should be treated separately. Serious injuries should be sent to a unit capable of dealing with them at any time of the day or night, renamed an 'accident and emergency' unit.

Accident and emergency departments would deal with seriously injured patients in urgent need of skilled hospital treatment. Continuous 24-hour cover meant that each unit must have at least three consultant surgeons, supported by adequate numbers of junior medical staff, nurses and auxiliary staff such as radiographers. This requirement imposed a lower limit of size on accident and emergency departments and consequently a minimum catchment of population, estimated at 150,000. It meant greatly reducing the number of hospitals offering an accident service and fitted in well with district general hospital policy. In the committee's view, it is normally better for a seriously injured patient to have a longer ambulance journey to a fully staffed accident and emergency department than to go to the nearest hospital, which may sometimes be able to give only first-aid treatment before sending the patient on to a larger hospital. No evidence was offered in support of this view. Neither was there discussion of any detrimental effects of increasing the travelling distances and times to accident units, although the committee acknowledged that in remote rural areas cottage hospitals should continue to give first-aid treatment to major injuries before sending them on. It seems that staffing considerations were given much more attention than the geographical implications of the policy.

The Platt Report was less clear about what to do with the patients with less serious injuries who had made up the majority (as much as 88 per cent) of people attending the former casualty departments. It envisaged a steady reduction in these patients because of the increase in group practices in which one practitioner is always available for emergency calls (perhaps an optimistic forecast). It acknowledged, however, that there will always be a demand for the treatment of minor injuries at hospital, for those injured at work or on holiday, for those whose general practitioner is unavailable and also for those who just arrive expecting treatment, 'despite all the efforts of general practitioners and hospital staff to discourage

Figure 4.1: The distribution of accident and emergency departments

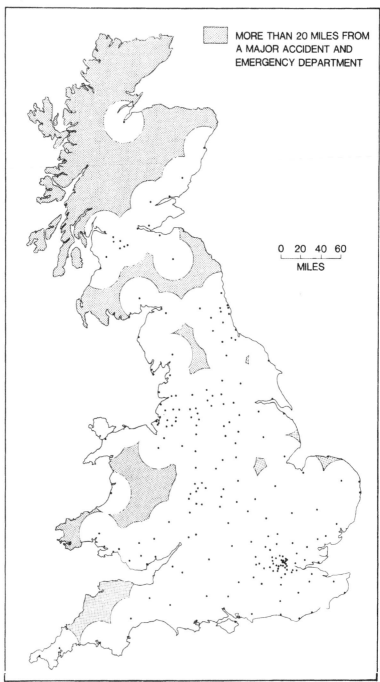

MORE THAN 20 MILES FROM
A MAJOR ACCIDENT AND
EMERGENCY DEPARTMENT

0   20   40   60
MILES

them' (DHSS, 1962, p. 26). The report recommended that separate provision should be made for patients with minor injuries in hospitals with accident and emergency units 'at least for the time being'. Ideally these should be staffed by general practitioners. Some other hospitals (including cottage hospitals in remote rural areas) not designated as accident and emergency centres might also provide treatment for minor injuries.

The arguments used in the Platt Report continue to be the foundations of accident and emergency planning. A report by the Expenditure Committee of the House of Commons and the government's response to it confirmed the policy of concentrating services because the advantages of large units were considered to outweigh the problems of long journeys (House of Commons, 1974, 1975). In 1970 there were 265 major accident and emergency departments. Less than 200 major departments were left by 1982 (Institute of Health Service Administrators, 1984), but many small local hospitals have continued to provide a minor casualty service. The distribution of major accident and emergency units is shown in Figure 4.1. Almost the whole of England is within 20 miles of a unit, with the exception of parts of north Devon, the northern Pennines and the Northumbrian coast. Much of central Wales and the Pembrokeshire coast is more than 20 miles from an accident and emergency department. In southern Scotland, large parts of the Borders Region and Dumfries and Galloway lie outside this distance. In northern Scotland much greater distances are involved, with the Scottish islands in by far the worst position.

As major accident and emergency facilities have been concentrating, accident and emergency attendance rates have been rising (Cartwright and Anderson, 1981). Typically, more than half the people who attend are not the types of patient for which the units were designed. A survey of six accident and emergency departments in and around London showed that just over one-third of all new patients attending had conditions that required hospital treatment. About 40 per cent of all the patients attending said they were not registered with a local general practitioner (Farmer, 1984). While some people go to accident and emergency units because they are not registered with a general practitioner (particularly in London, where non-registration is exceptionally high), other 'inappropriate' attenders do have their own doctor. Dennis (1984) has reported the reasons people give for using a hospital service rather than their own family doctor. The availability of a 24-hour service in hospitals is well known. The impersonal institutional atmosphere of a hospital may be preferred by some people to their local doctor's crowded waiting room. Some people may be unable to telephone to make an appointment with their general practitioner, either through lack of access to a telephone or through insufficient ability in the language. Some may not have changed their general practitioner since moving from somewhere else. Others feel that their own doctor has not understood the problem or would not wish to be bothered with it.

Farmer (1984) points out that for many people the accident and emergency unit is a natural source of primary medical care. Their preference is inconvenient, but it is also real. They account for about half the workload of most units. Refusing access to accident and emergency services and directing people to a local general practitioner instead would be difficult in practical terms (the decision of who to turn away would have to be made by a doctor) as well as introducing ethical problems. If access could be inhibited, workload would decrease substantially but costs would not fall in the same proportion. Accident and emergency centres would still need to maintain full 24-hour cover.

Psychiatric hospitals

The movement started by the Hospital Plan in 1962 to reduce the number and size of psychiatric hospitals soon gained support. Most large psychiatric hospitals were founded in the last century when they were expected to perform a custodial role. As a result many people who were mentally ill or mentally handicapped were confined in hospitals for social rather than medical reasons. Their treatment as 'patients' was seen as inappropriate in many instances. Under the hospital plan, psychiatric hospitals were to be gradually run.down, and would ultimately cater for a much smaller proportion of people with identified medical or nursing needs.

Considerable reductions in the psychiatric hospital service followed. Figure 4.2 shows the trends for mental handicap hospitals in the South Western Region, with all hospitals with over 250 beds in 1971 being scaled down in subsequent years,

104

Figure 4.2: The rundown of the larger mental handicap hospitals in the South Western Region

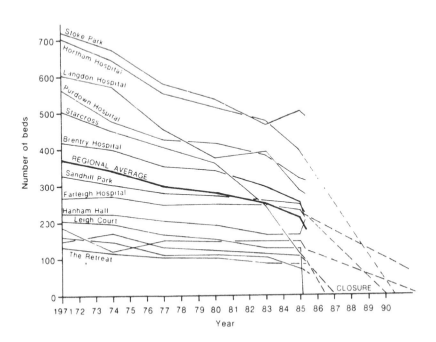

Source: Phillips and Radford, 1985, p. 10.

and some being identified for closure. Former patients have been transferred mostly to smaller residences administered by social services departments (DHSS, 1971a). Because the hospital system for mentally ill or mentally handicapped people was centralised in relatively few, large institutions, transferring patients to their home areas frequently involves movements across health district boundaries or even regional health authority boundaries, in which case funds are also transferred in the same direction. In the Frenchay and South-mead Districts, for example, 41 per cent of residents either originated from or have their nearest living relative outside the South Western Region (Phillips and Radford, 1985). For elderly residents, it may be a problem simply to identify their most appropriate home area.

Mental illness hospitals have followed similar trends, with more reductions than actual closures to date. The Royal

105

Commission (1979) noted that although 13 mental hospitals in England and Wales with over 300 beds each were expected to close in the 1960s and 1970s, in fact only one did so − and this was converted into a hospital for mentally handicapped patients. Alternative provision for mentally ill patients has not materialised in district general hospitals as was planned in 1962, and the Royal Commission detected no sign of a movement in this direction in the foreseeable future.

The policy of discharging mentally ill and mentally handicapped people into the community relies upon being able to provide additional support for them outside hospital. The main source of this extra support is 'joint funding', which allows NHS money to be used by local authority social services departments to make alternative provision. Somerset's 'core and cluster' scheme (Phillips and Radford, 1985) is an example of a plan to transfer mentally handicapped people into the care of a social services department through the medium of joint funding. In this model a 'core' building for administration, short-term care and crisis care is surrounded by a cluster of houses, each with staff appropriate for the residents' level of dependency. Other districts in the same region have less radical plans which may involve using ordinary housing as much as possible and relying for support on conventional domiciliary services such as district nurses, social services, home helps or meals-on-wheels. Responses differ between districts, but Phillips and Radford report an overall slow but steady progress in the South Western Region towards the goal of offering mentally handicapped people as normal a life as possible outside hospital.

Nationally, however, and particularly in the context of mentally ill people, the widespread reduction in beds in psychiatric hospitals has been interpreted as a money-saving tactic with little benefit to many of the people involved. Gibbons (1978), for example, has argued that families now bear an increasing burden of the social costs involved. Many ex-patients not supported by their families live in boarding houses. The least fortunate are sleeping on the streets or are in prison. The House of Commons Social Services Committee (1985) heard disturbing evidence from across the country and came to the conclusion that the rundown of psychiatric hospitals is not being matched by the establishment of adequate community services. It recommended that no further reductions should be made in hospital places without providing sufficient alternatives in advance. The problem is not wholly financial, of course. Plans to establish community psychiatric facilities are often blocked by organised opposition from local residents who fear negative effects on themselves.

Community hospitals policy

Early hospital plans required the closure of most small hospitals but, in the event, this proved to be an impractical and unpopular policy. The only coherent national plan to make use of small hospitals in a role different to that assigned to the district general hospital was known as the community hospital policy (DHSS, 1974). Community hospitals were defined as units providing medical and nursing care, including out-patient, day patient, and in-patient care, for people who did not need the specialised facilities of the district general hospital. They were envisaged as having 50–150 beds, which meant that they would be larger than one-third of existing hospitals at the time. Community hospitals were not intended to increase the total number of beds planned to be provided per thousand population, but could be used instead of the district general hospital for some geriatric patients, elderly severely mentally infirm patients and a third category of medical and surgical patients. This included preconvalescent patients transferred from the district general hospital after the most intensive part of their treatment, people with terminal illness and people with chronic disabilities who needed nursing care to relieve their families temporarily. The daily care of patients in community hospitals would be the responsibility of general practitioners, who ideally should have their practice premises or health centres on the same site. Day hospital facilities would be provided for geriatric patients and for elderly mentally infirm people. It was also intended that consultants should hold appropriate out-patient clinics at community hospitals.

Although the 1974 circular recognised the need for local flexibility, it did attempt to establish a standard model for smaller hospitals as a second layer in a new two-tier hospital hierarchy. This was not successful. There were widespread misgivings about some of the guidelines, particularly those concerned with elderly severely mentally infirm patients

107

(Royal Commissión, 1979). Furthermore, many smaller hospitals were still needed to fill gaps in the general hospital service. In the West Norfolk and Wisbech District, for example, relatively few of the in-patients in hospitals were considered by consultants to be suitable for care in units supervised by general practitioners (Haynes and Bentham, 1979b). With the general hospital service working at full stretch, implementation of the community hospital policy required additional resources that were not available. As little as three years later the Department of Health and Social Security acknowledged that a widespread new building pro-gramme for community hospitals was unlikely. The concept of a community hospital was widened to include any retained local hospital which was not part of a district general hospital service. Health authorities were advised that the guidelines should not stand in the way of practical local solutions (DHSS, 1977).

In the event, the practical local solution was often to retain small hospitals as necessary annexes of the general hospital or to replace them by new district general hospital facilities. Most general practitioner 'cottage hospitals' were kept as they were. There has, however, been a significant increase in day care in smaller hospital units. There are no official figures, but an estimate was made of 217–302 day hospitals for elderly people in Britain in 1977, with most of them opened since 1970. Current provision was given as about one place per 1,000 elderly people (DHSS, 1981c).

General practitioner hospitals

The traditional 'cottage hospital' run by general practitioners for the benefit of their patients has been little affected by community hospital policy, although the national policy of concentrating maternity services in district general hospitals for safety reasons has had an impact. Cavenagh (1978) estim-ated that there were about 350 general practitioner hospitals in England and Wales. He conducted a one in seven survey of them and found that the sample had an average of 24 beds. Their average distance from the nearest district general hospi-tal was 22 kilometres. Most of them were visited by consul-tants and had physiotherapy and X-ray facilities. Almost all

108

treated casualties and 99 per cent of casualties who arrived at the hospital received all their treatment there (serious accident and emergency cases were usually taken directly to a district general hospital by ambulance crews). Nearly half the hospitals undertook surgery (more by consultants than general practitioners). Surgery was limited in range but large in volume. Cavenagh estimated that general practitioner hospitals were dealing with 6 per cent of all operations, 13 per cent of all casualties and 4 per cent of X-ray examinations in England and Wales. He calculated that 20 new district general hospitals would be needed to cope with the same workload: arguably a more costly option for the NHS and less convenient for patients. Cavenagh, a general practitioner himself, is an advocate of expanding the general practitioner hospital service.

In Scotland, Grant (1984) identified 64 general practitioner units, containing 3 per cent of Scottish hospital beds. Although this proportion is low, and most general practitioner units in Scotland have only 10–20 beds, Grant argues that a significant contribution is made. The majority of Scottish general practitioner hospitals are more than 30 miles from the nearest district general hospital and serve rural communities. Their beds are used for acute conditions, long-stay patients, maternity patients and chiefly minor surgery cases, in that order of frequency. General practitioner maternity services (not directly supervised by consultants) were provided in 35 of the 64 units, and 7 of them were exclusively obstetric hospitals. They accounted for over 4 per cent of the total number of deliveries in Scotland in the study year.

It is in the facilities other than beds for in-patients that the scope of Scottish general practioner hospitals may be assessed. Simple X-ray facilities were available in 59 per cent of units, physiotherapy in 69 per cent, and 19 per cent had day centres for the elderly. In most hospitals (86 per cent) consultant out-patient clinics were provided in the major specialties at least once and sometimes twice a month. All of them, with the exception only of the specialised maternity units, provided a general practitioner casualty service.

In defence of general practitioner hospitals in Scotland, Grant stresses the relief of pressure they bring to the acute beds in the district general hospital, the relief of transport difficulties for 140,000 out-patients and 100,000 casualties per year, the advantages for recovery and rehabilitation of elderly patients cared for by their own doctor within easy reach of their families, and the employment of local nurses, physiotherapists and radiographers whose skills might otherwise not be used. Neither he nor Cavenagh considered the possible disadvantages of general practitioner hospitals, which include the diseconomies of very small size and the dangers of attempting some forms of treatment with inadequate resources.

Hospitals in transition

Between May 1979 and December 1983, 112 units with an average of 91 beds were closed. During the same period, 31 new units with an average of 381 beds were completed (House of Commons, 1984). The majority of the new units were district general hospitals or the first phase of district general hospitals. The pattern of hospital services is currently in flux. It is still far from being an orderly system of large district general hospitals supported by a network of smaller units with different, complementary functions. Small acute hospitals are closed as they are gradually replaced by larger general hospitals, while large psychiatric hospitals are being run down as domiciliary and day centre services are provided, in theory at least, to fill the gap. This process, which is guided by policy, is complicated by the 'cuts': reductions in hospital funding which have mostly affected health authorities in the four Thames regions. In London, savings in revenue expenditure have been made by closing hospitals without replacement (Chapter 3).

Hospital sizes are still spread over a very wide range, even though the trends for non-psychiatric hospitals to become larger and for psychiatric hospitals to become smaller are marked (see Table 4.1). In a 1983 listing, the largest non-psychiatric hospital is St James' University Hospital in Leeds, with 1,424 beds (Institute of Health Service Administrators, 1984). At the other end of the scale, more than one-third of the non-psychiatric hospitals in England remain smaller than the minimum size of 50 beds recommended by the DHSS and 15 per cent are smaller than 25 beds. Small non-psychiatric hospitals are especially numerous in the South Western Region, where 56 per cent have less than 50 beds and 32 per cent of hospitals are smaller than 25 beds. For psychiatric

Table 4.1: Hospital sizes in England

Hospital type and size	1959	1980
Non-psychiatric hospitals:		
with under 50 beds	912	575
50–249 beds	925	690
250–499 beds	216	180
500–999 beds	76	105
1,000–1,999 beds	9	10
Total	2,138	1,560
Psychiatric hospitals:		
with under 50 beds	65	152
50–249 beds	86	112
250–499 beds	27	59
500–999 beds	43	75
1,000–1,999 beds	54	26
2,000 and over beds	28	—
Total	303	424

Source: DHSS, 1982, p. 61.

hospitals, there have been dramatic reductions in size, although some very large institutions persist. Winwick Hospital near Warrington had the largest number of beds (an unenviable 1,469) in the 1983 list.

The plan for the Blackburn Health District, described by Grime and Whitelegg (1982) is an example of how an inherited pattern of hospitals is being reorganised to fit national policy as closely as local circumstances permit. The Blackburn District has a population of some 270,000, mostly in the urban areas of Blackburn and Accrington in the south. The Ribble Valley area to the north is rural and relatively sparsely populated. Before reorganisation (see Figure 4.3), most of the district's in-patient specialties were provided in two large hospitals in Blackburn. A smaller hospital in Accrington had some acute specialties and other small units scattered over the district provided notably geriatric and general practitioner maternity beds and some other specialised beds. The main aim of reorganisation was to concentrate most acute specialties into a single district general hospital, Queen's Park Hospital, on a large site. Bradford Royal Infirmary consequently is to lose some of its acute beds to the new district general hospital, and Accrington Victoria Hospital will lose all of them. Accrington Victoria Hospital, like the other smaller

111

Figure 4.3: Changes in hospital provision in the Blackburn
health district: Services in 1979 and those planned for 1987/8

4.3a

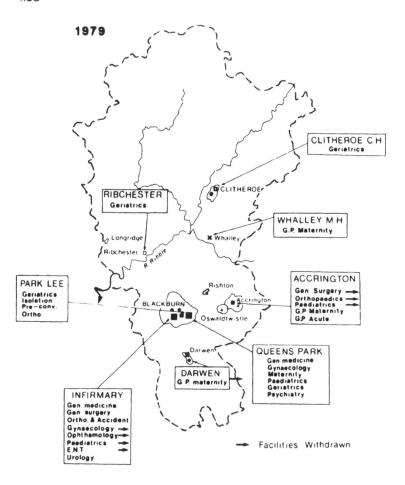

Source: Grime and Whitelegg, 1982, pp. 204–5.

units, is to take on some of the functions of a community
hospital, with beds for geriatric patients, the elderly severely
mentally infirm, general practitioner medicine and (elsewhere)
for preconvalescent patients. Only one general practitioner
maternity unit is planned to be retained. The new pattern is
due to be completed in 1988. The bare facts of these changes,

Figure 4.3— *cont.*

4.3b

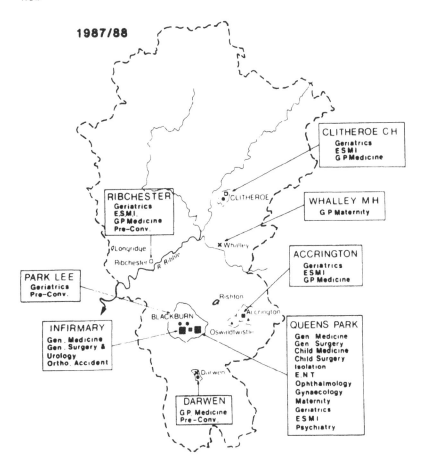

Source: As Figure 4.3a.

or the changes in any other health district, do not convey the deep local controversies involved, the disruptions to established working patterns or the attempts by interested groups to save particular institutions. Rationalisation has nowhere been an easy process.

CENTRALISATION VERSUS DISPERSAL

There are strong arguments for concentrating hospital services. Almost all the DHSS reports on particular services – general hospital services, operating facilities, accident and emergency facilities, maternity services and so on – supported the district general hospital concept. Yet the case was made from the point of view of increasing efficiency and effectiveness within the NHS. The welfare of patients was considered in terms of medical outcome. Matters such as the organisation of work, patterns of staffing, career structures and the cost of technological equipment generally received the most detailed scrutiny. In contrast, little attention was given to the broader interests of the public at large. The non-medical costs and benefits to the community that centralisation implied received scant mention.

Locally, however, the wider issues cannot be evaded. The closure of small local hospitals continues to be a lively political issue. Raising the particular case of Bretby Hall Orthopaedic Hospital in South Derbyshire, which was to be replaced by the district general hospital in Derby, the Conservative Member of Parliament for South Derbyshire recalled more than 100 Commons questions and eight adjournment debates about hospital closures in the previous year (Currie, 1984). Many politicians have responded to the strength of local opinion and have championed the cause of keeping small hospitals open. Currie puts forward the view that new hospital buildings are more expensive to run and less attractive for both patients and staff than the older buildings they are designed to replace. Small hospitals are often generously supported by the local community and the new larger units are invariably located in large towns and cities, which are losing population, rather than in the suburbs and rural areas, which are gaining. The last argument is a double-edged weapon, of course, since a new urban site may in fact be accessible to more rural people than a location out of town (as the health authority claimed was the situation in the Bretby Hall controversy).

Small hospitals for geriatric, psychiatric and mentally subnormal patients are quite commonly in inconvenient locations because they occupy former workhouses and hospitals administered by the Poor Law authorities, which were originally built on cheap land at some distance from the centres of population. The sites of general hospitals are likely to be accessible for many more people than this type of location. To illustrate the point, McKeown *et al.* (1971) compared visiting to geriatric and psychiatric patients in centrally located district general hospitals in Birmingham with visiting to similar patients in smaller units, mostly situated outside the city. They found that the duration of travel was about a third less for visitors to patients in district general hospitals than for those who visited patients in specialised outlying units. Furthermore, visiting rates were higher in the district general hospitals. They concluded that visiting conditions to geriatric, psychiatric and mentally subnormal patients in Birmingham would be considerably improved if all such services were provided from district general hospitals.

Not all small hospitals are accessible to their patients and many do not attract the enthusiastic support of the local community. Hospitals for geriatric, psychiatric or mentally subnormal patients are rarely defended as vigorously as small general hospitals or maternity units. Political resistance is more likely to be aroused when a hospital catering for acute conditions in an identifiable small catchment area is threatened. But public pressure is unlikely to overcome the weight of medical opinion that the majority of acute hospital services, including maternity services, are safer, more effective and cheaper when provided in large centralised units. The most promising future for the small local hospital is in supplying the minority of services to which the case for centralisation does not apply. Although the 'community hospital' initiative ran out of steam, the arguments behind it remain valid and are worth rehearsing.

Local in-patient services

The majority of hospital patients are admitted for a short period while acutely ill or to undergo surgery. They require the facilities of a district general hospital and any inconvenience in getting to the hospital for patients and visitors is considered tolerable because of the rarity of the experience and the short time periods involved. There are other people currently in hospital, however, who do not need expensive investigations or sophisticated forms of treatment but who

115

require longer-term nursing or rehabilitative support which cannot be provided at home. Many elderly people in hospital, the chronically sick and some severely physically or mentally handicapped people belong in this category. Elderly people in particular might be alarmed by the clinical atmosphere of a large general hospital. Where possible, a more domestic environment on a less institutional pattern, staffed by familiar people and especially with the patient's own general practitioner at hand, is desirable. Continuity of care by the general practitioner is also an acknowledged medical benefit.

In-patient services in small local hospitals are perhaps best viewed not as extensions of the district general hospital but as support for care at home. Many people who are looked after at home might be helped by a short period of rehabilitation in a local unit that would prolong their lives in the community and temporarily ease the burden on their relatives. Other people who could benefit from physiotherapy, occupational therapy or alternative remedial treatments might receive such help in the day ward of a local hospital, returning home in the evening. In this way, small local hospitals might help to foster greater contact and co-ordination between the health services, social services and the public. A combination of domiciliary care backed up by local hospital support is an attractive alternative to long-term institutionalisation.

Such advantages depend on the small hospital having a local catchment area which makes it easy to reach by patients and their visitors, so that use of its facilities can be part of day-to-day life rather than a disruption to it. Small hospitals with a local identity are a focus of community interest and pride (never so evident as when closure is threatened) and a source of energetic voluntary activity. Local job opportunities may also be attractive to trained staff living outside the main urban centres who perhaps might not wish to travel to work in a more distant general hospital. This particular argument is not a strong one since Sadler and Whitworth (1975) found in a national survey of nurses that marriage and children (not distance) are the main factors keeping trained nurses out of work. Furthermore, the key groups of radiographers, physiotherapists and occupational therapists, who are also in short supply, may be more attracted by the professional opportunities in large hospitals than the domestic atmosphere of a local unit (Haynes and Bentham, 1979b), so recruitment would not necessarily be easy.

Medical staff, on the other hand, would almost certainly be available for small local hospitals on a modified community hospital pattern, given acceptable terms and conditions. General practitioners would be more appropriate than hospital consultants to provide the day-to-day care of the types of in-patient envisaged, and many general practitioners would welcome increased access to hospital beds. A survey of all general practitioners in the West Norfolk and Wisbech District (Haynes and Bentham, 1979b) uncovered a high level of interest in active participation in a small local hospital. A large majority said they would be interested in working in a community-type hospital up to five miles from their home. As many as 57 per cent of general practitioners outside the towns said they would be willing to travel five to ten miles to work in such a hospital.

There remains the question of how small a local in-patient unit could be, while still remaining viable in terms of cost per bed. Rickard (1976a) analysed data on the average revenue costs of about 500 hospitals with less than 100 beds. A quadratic function was found to best describe the relationship between average cost per bed and the number of beds. Very small hospitals had high average costs per bed, but as size increased average costs fell, reaching a minimum at 35 beds. Above that size, there was little variation in average running costs with size. This is an important finding if it is generally applicable, because it suggests that a hospital unit with a restricted range of facilities might be as small as 35 beds without losing economies of scale. From the consumer's point of view, two small hospitals each with 40 beds would almost certainly imply greater social benefits than a single 'small' hospital with 80 beds.

Dispersed out-patient clinics

A second major group of hospital services which might be devolved to the local level with advantages to the community and little cost to the NHS is that of out-patient clinics. Although some patients need the specialised diagnostic and treatment facilities of a district general hospital when they attend an out-patient clinic, others could be examined by a consultant satisfactorily in a small hospital, a health centre or a local community health clinic. When out-patient clinics are dispersed, the main burden of·travel falls on hospital consultants rather than patients.

Gruer's (1972) study of out-patient referrals in the Borders Region of Scotland shows how a mixed system of out-patient clinics operates. In this rural area, general practitioners could choose between three types of clinic when referring patients to a consultant: simple clinics in very small cottage hospitals visited by consultants from elsewhere, clinics in a small general hospital serving the region and specialised clinics in more distant Edinburgh hospitals. Gruer found that the process of choice was likely to be complex, but the two main considerations were distance (general practitioners appeared to attempt to minimise patient travel) and personal contacts with hospital consultant staff (general practitioners would refer patients to former undergraduate colleagues, for example). In the event, 48 per cent of all new out-patient referrals were to the minimally equipped cottage hospital clinics. Some of the differences between these patients, who were able to take advantage of using a more local clinic, and the patients who were sent either to the region's general hospital or to Edinburgh may be detected from Table 4.2.

Table 4.2 makes it clear that the main reason for referral to any out-patient clinic was for the patient to receive consultant advice, but especially for cottage hospital referrals. The lack of specialised treatment facilities in cottage hospital clinics is reflected in the lower percentage of patients referred for treatment compared with the general hospital referrals. Although few diagnostic aids were available in the cottage hospital clinics, a similar proportion of definitive diagnoses was made there compared with the more specialised clinics, possibly because most patients referred to the simpler clinics had common and uncomplicated conditions. The majority of these conditions would probably require only a confirmation of diagnosis by clinical examination and subsequent placement on a waiting list for admission. But, as Gruer points out, the same was also true of referrals to Edinburgh hospital clinics, so the difference is just one of degree. After the first visit to the clinic, there were variations in the disposal of patients attending the three types of clinic, with a higher proportion of

Table 4.2: Characteristics of out-patients referred to Edinburgh hospitals, the local general hospital or cottage hospital clinics in the Scottish Borders Region (%)

Characteristic	Edinburgh	General hospital	Cottage hospital
Reason for referral:			
Consultant advice	84	82	96
Treatment	13	17	3
Investigation	2	< 1	< 1
Patient reassurance	1	< 1	< 1
Level of diagnosis:			
Firm	64	66	68
Provisional	18	26	18
No diagnosis	8	4	8
Nothing wrong	9	4	6
Disposal:			
Admit to hospital	3	7	2
Hospital waiting list	19	20	42
Further investigations	8	12	8
Further appointment, same hospital	37	24	18
Further appointment, different hospital	0	14	8
Return to GP	32	21	21
Base	225	687	1,072

Source: Gruer, 1972, pp. 23–6.

cottage hospital out-patients being put on waiting lists for admission as in-patients (mostly for repair operations) and a lower proportion continuing as out-patients. Only about 8 per cent of cottage hospital out-patients were referred to a more specialised clinic at another hospital. This suggests that the cottage hospital clinics' facilities were inadequate for only a very small proportion of patients referred there. If these patients had been referred elsewhere – to the general hospital, for example – some of them might still have been given further appointments (38 per cent of patients at the general hospital were). Some of the 8 per cent, however, may have been patients who could have made a satisfactory single trip to a more specialised clinic. For these people, and for the consultants who saw them, referral in the first instance to a simple local clinic was probably a hindrance rather than a help.

The widespread practice within the NHS of concentrating out-patient clinics in well-equipped general hospitals has evolved under a planning system which considers the internal

119

costs to the health service, but not the overall costs (both social and economic) to the nation. If more out-patient clinics could be held locally, outside the district general hospital, the reduced amounts of travel and working time lost by patients would numerically far outweigh the increased travel time and costs for consultants, as Gruer was able to demonstrate. Small local clinics are less intimidating for patients (but also less anonymous, which may sometimes be a disadvantage). There are benefits in terms of both continuity of care and mutual education if the general practitioner and the consultant are in contact at the time of referral. On the other hand, contact between hospital doctors themselves, their access to information services such as libraries and laboratories and their availability for emergencies at the main hospital all tend to suffer. Working with incomplete facilities in peripheral clinics is a real difficulty in some specialties, but by no means all.

A report which considered that dispersed out-patient clinics could be held in health centre premises suggested that the most suitable specialties are general medicine, dermatology, psychiatry, paediatrics, obstetrics and some aspects of geriatrics (DHSS, 1971b). To be suitable for devolution, a specialty must have both a large flow of out-patients and a large proportion who do not need specialised diagnostic facilities. Haynes and Bentham (1979b) asked all hospital doctors in one health district to estimate what proportion of their own out-patients could be seen satisfactorily in a small local hospital with simple facilities. Estimates varied between specialties, with the highest proportions in psychiatry, general surgery, general medicine, obstetrics/gynaecology and paediatrics (all 75 per cent or more). Altogether 51 per cent of the total first attendances in the district and 57 per cent of reattendances were thought to be suitable for peripheral out-patient clinics. This tallies with Gruer's evidence, which suggested that up to 50 per cent of all new out-patients do not need the facilities of a centralised clinic. The case for retaining some specialised clinics in district general hospitals is not at issue, but there is now evidence to support a strong argument for a system of local clinics with general practitioners selecting the most appropriate destination for each patient.

Comparing locational strategies

Any plan to alter the geographical distribution of hospital services has an effect on the accessibility of the service to the public and the travel costs incurred by the community. Plans to extend a hospital or to close another are always accompanied by detailed estimates of the costs to the National Health Service, but rarely by an equivalent exercise from the point of view of the consumer. Gruer (1972) drew attention to the one-sidedness of this approach and, using the example of out-patient clinics, demonstrated that an appraisal which balanced the costs of the community against those of the health service was perfectly feasible.

Gruer estimated the annual costs to the NHS of hospital consultants attending local clinics in the Borders Region of Scotland by adding the cost of consulting time lost through travelling to the actual mileage expenses. The costs to the community in travelling to out-patient clinics in cottage hospitals, the general hospital and hospitals in Edinburgh were calculated by adding estimates of loss of earnings to travel expenses. This exercise involved making assumptions about the loss of time for out-patient trips, average wages, female employment rates, modes of transport used, and so on. The existing situation, with out-patient clinics being attended in cottage hospitals, the region's general hospital and in hospitals in Edinburgh, was then compared with several other (hypothetical) arrangements of clinics in the same region, of which two appear in Table 4.3. The first of these (Model B) was the closure of all peripheral cottage hospital clinics, with the patients who would have been referred to them being directed instead to the region's general hospital. The second (Model C) incorporates the plan for a new, larger, district general hospital to replace the existing general hospital at a different and more accessible location. The cottage hospital clinics are retained in this arrangement and the increased specialisation of the new district hospital makes it possible to reduce the use of Edinburgh clinics.

Table 4.3 shows clearly that the travel costs incurred by the health service are modest compared with those borne by the community in all three models. The potential savings to the NHS in travel costs which would result from closing the small peripheral clinics are strongly outweighed by increased costs to

Table 4.3: Estimates of annual consultant and community travel costs for alternative arrangements of out-patient clinics in the Scottish Borders Region (£)

Model	Consultants' costs	Patients' costs	Total travel cost
A. Existing situation: Local general hospital, peripheral clinics, some use of Edinburgh clinics	6,700	17,700	24,400
B. Closure of peripheral clinics: Local general hospital, some use of Edinburgh clinics	4,200	23,300	27,500
C. Replacement of hospital by DGH: New DGH, peripheral clinics, reduced use of Edinburgh clinics	3,200	15,400	18,600

Source: Gruer, 1972, pp. 43–7.

the community. The most economical model for both consultants' and patients' travelling costs is that of Model C, which incorporates the advantages of both a centrally located and well appointed district general hospital with a network of peripheral clinics with basic facilities.

Compared with the high costs of medical care, the costs involved in consultant travel are small indeed. As Gruer argues, for a modest outlay on consultants travelling to cottage hospitals, there can be a striking reduction in the distance travelled by patients. There can be no doubt that a reduction in travel time for people who are unwell and a reduction in travel cost for those living on low incomes, sickness benefit or old age pensions are both desirable goals.

Other researchers have since repeated similar modelling exercises which support Gruer's claim that the community's costs of access to hospitals vary substantially between different locational strategies and can be estimated accurately enough for comparative purposes. Haynes and Bentham (1979c) also studied different deployments of out-patient clinics, in the West Norfolk and Wisbech District of East Anglia. Figure 4.4 shows four possible distributions. Strategy B was the existing arrangement, with clinics held in hospitals in King's Lynn and Wisbech. Strategy A was the most centralised pattern possible in the district, while C and D represented possible future dispersal strategies. The objective was to measure annual

Figure 4.4: Possible locations and catchment areas of out-patient clinics in the West Norfolk and Wisbech District

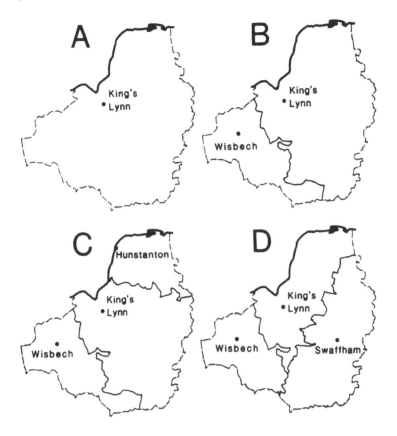

Source: Haynes and Bentham, 1979c, p. 119.

expenditure on transport for each of the four arrangements. First, the annual number of clinic attendances expected to originate from each parish or town was calculated by distributing the current 33,000 annual attendances over the district in proportion to parish populations (age and sex weighted). People were assumed to attend the nearest clinic, so the final costs were likely to be underestimates. For each trip, the mode of transport, the time taken and the cost were all calculated on the basis of known travel patterns in the study area. Aggregate results are given in Table 4.4, which shows how transport costs for the community and likely ambulance mileage for the NHS are reduced from the most centralised strategy (A) to the

123

Table 4.4: Transport implications of four locational strategies for out-patient clinics

Strategies	A	B	C	D
All attendances:				
Mean distance (miles)	9.9	7.4	6.4	5.8
Total miles (thousands)	657	490	425	381
Total hours (thousands)	34.9	29.4	27.1	25.5
Percentage over 10 miles	56	33	25	22
NHS transport:				
Mean miles per patient	23.2	18.5	16.5	14.8
Total miles (thousands)	116	80	66	59
Community transport:				
Mean miles per patient	19.1	14.1	12.3	11.0
Total miles (thousands)	541	410	359	322
Mean cost (£)	1.05	0.79	0.69	0.63
Total cost (£ thousands)	35.1	26.4	23.1	20.9

Source: Haynes and Bentham, 1979c, p. 119.

most accessible dispersed strategy (D). The detailed cost and time estimates were found to be unnecessarily complicated when comparing the accessibility implications of the various strategies. The average distance to hospital was a simple summary measure which closely matched variations in cost and time estimates.

Grime and Whitelegg (1982) have also demonstrated that it is possible to measure the accessibility implications to consumers of hospital location plans. They modelled travel to hospitals before and after the implementation of a strategic plan to reorganise the hospital service in the Blackburn Health District. The essence of the plan was to concentrate most acute specialties at a single site (Queen's Park Hospital in Blackburn) while allowing dispersal of geriatric, elderly severely mentally infirm and some medical beds to smaller local hospitals. The effects of the plan were therefore to increase patient and visitor travelling distances for most acute specialties while reducing them for other specialties. Grime and Whitelegg were able to estimate the magnitude of these changes and to conclude, for example, that substantial savings would be made by visitors to geriatric patients, even under a wide range of assumptions about visiting frequencies. More exercises of a similar type are needed, to demonstrate to health authorities that it is practicable as well as desirable to evaluate the effects of their policies on consumers.

Table 4.5: Regional variations in the average number of hospital beds available per thousand population, 1980

	All special- ties	Acute special- ties	Medical	Surgical	Mental illness	Mental handi- cap	Geriatric[a]
England	7.7	2.8	1.1	1.7	1.9	1.0	7.8
Northern	8.2	3.2	1.2	1.9	2.0	1.1	8.4
Yorkshire	8.2	3.1	1.2	1.9	2.0	1.0	9.7
Trent	7.0	2.4	0.9	1.6	1.7	1.0	7.7
East Anglian	7.0	2.5	0.9	1.6	1.6	0.8	8.4
NW Thames	7.8	2.8	1.2	1.6	2.3	1.1	6.6
NE Thames	7.8	3.2	1.4	1.8	1.9	0.7	7.8
SE Thames	7.8	3.0	1.2	1.8	2.0	0.9	6.9
SW Thames	8.5	2.3	0.8	1.5	2.5	2.0	6.2
Wessex	6.9	2.4	0.9	1.5	1.5	0.9	8.1
Oxford	5.8	2.2	0.9	1.3	1.1	0.9	7.3
South Western	8.0	2.4	0.8	1.6	1.9	1.6	7.1
West Midlands	6.9	2.6	1.0	1.6	1.6	0.8	8.8
Mersey	8.7	3.2	1.3	1.9	2.6	1.0	9.0
North Western	7.6	3.2	1.2	1.9	1.7	1.0	8.0

Note: a. Geriatric figures are beds per 1,000 population aged 65 and over.

Source: DHSS, 1982, p. 71.

REGIONAL DISPARITIES

Table 4.5 shows that there are still notable differences between regions in the provision of hospital beds. Data such as these should be interpreted carefully because the regions with relatively low rates of bed availability are not necessarily providing the worst care. Hospital beds for mental illness and mental handicap, for example, are being reduced as a matter of national priority, so regions with low proportions of these beds could either be decades behind the rest of the country in building up services or, on the other hand, they may be in the forefront of pioneering better alternatives. Acute medical and surgical specialties, on the other hand, are generally acknowledged to be most appropriately provided in hospitals. The acute specialty figures show that the Northern, NW Thames, Mersey and North Western regions have almost 50 per cent more beds available per thousand population than the Oxford Region. This does not allow for age and morbidity variations between regions, which would tend to reduce the discrepancy. Within regions there are greater differences, with a marked bias towards inner-city areas. In the NE Thames Region, for example, the City and East London Area had 5.12 beds per thousand population in 1975, compared with 2.11 beds per thousand population in the Essex Area. It was estimated that up to 25 per cent of all people admitted to beds in the City and East London Area were from outside its boundaries (Eyles and Woods, 1983).

When hospital beds are divided into groups according to the diagnostic condition for which they are used, even larger variations between regions are apparent. Table 4.6 gives the range from the highest to the lowest regional bed rates for the main categories of medical condition recognised by the World Health Organization (mental disorders and conditions of pregnancy and childbirth are omitted). The data refer to the period 1976–8. The most extreme difference was between the bed usage rates of 2.7 per 100,000 population in SE Thames and 10.4 per 100,000 in the North Western Region for infective and parasitic diseases. Several other conditions had bed usage rates with more than a twofold difference between regions. Only a small proportion of these variations appeared to be due to age and mortality risk differences between the regions. Standardising the bed rates to take account of regional variations in age structure and mortality altered the extreme rates for infective and parasitic diseases to 2.71 (SE Thames) and 9.56 (North Western), for example, but the basic spectrum of differences across the country was only slightly affected (Figure 4.5).

Hospital bed use rates are much more closely related to provision than to need (Cullis *et al.*, 1981). It is likely that the regions in Table 4.6 with low bed use rates for particular conditions are those which have low bed provision through the differing interests of consultants and the quirks of history. Similarly, high bed use probably indicates relatively generous provision in the past reinforced by inertia. If hospital beds are to be reduced in the future, in accordance with national policy, then the below-average regions will simply be a few years ahead of the rest. But not even the most charitable observer would claim that all the regional health authorities with low bed rates have been deliberately anticipating future priorities or that the present large and arbitrary variations from one region to another are a step towards future equitability. This exercise shows that the redistribution of resources

Table 4.6: Daily bed rates per 100,000 population: regional maxima and minima, by condition

Group no.	Condition	Maximum rate		Minimum rate	
—	All conditions	420	(North Western)	269	(Oxford)
I	Infective and parasitic diseases	10.4	(North Western)	2.7	(SE Thames)
II	Neoplasms	46.1	(North Western)	27.5	(West Midlands)
III	Endocrine diseases	14.2	(Wales)	7.7	(East Anglia)
IV	Diseases of blood	4.6	(NE Thames)	2.0	(Wessex)
VI	Diseases of nervous system	35.8	(South Western)	19.1	(Wessex)
VII	Diseases of circulatory system	102.0	(North Western)	46.6	(Oxford)
VIII	Diseases of respiratory system	33.0	(North Western)	15.4	(East Anglia)
IX	Diseases of digestive system	32.5	(North Western)	19.7	(East Anglia)
X	Diseases of genito-urinary system	21.7	(North Western)	11.6	(East Anglia)
XII	Diseases of skin	6.6	(Wales)	3.8	(East Anglia)
XIII	Diseases of musculo-skeletal system	27.0	(Yorkshire)	18.2	(Oxford)
XIV	Congenital anomalies	6.2	(North Western)	2.8	(East Anglia)
XV	Perinatal morbidity	5.0	(NW Thames)	2.4	(East Anglia)
XVI	Ill-defined conditions	48.8	(SE Thames)	25.1	(North Western)
XVII	Accidents and other injuries	40.1	(Wales)	23.3	(Oxford)

Source: Haynes, 1985, p. 22.

Figure 4.5: Regional bed usage rates for infective and parasitic diseases, with age and mortality adjustments

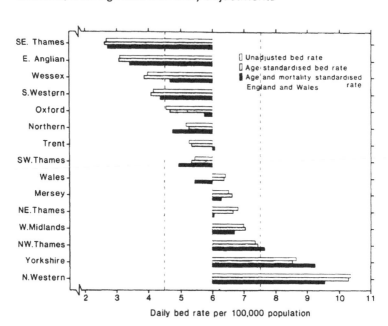

Source: Haynes, 1985, p. 25.

recommended by RAWP may not be appropriate across all non-psychiatric in-patient services. The London and Oxford regions are not the most generous users of hospital beds for every type of illness and the northern and western regions are not necessarily relatively deprived: it depends on the diagnostic condition. Differences at the district level are likely to be even more pronounced.

In Scotland the most pronounced disparity in hospital provision was between the four cities, notably Edinburgh and Glasgow, where most acute hospitals were concentrated and the remainder of the country. This required considerable travel from outlying districts to the cities for general as well as regional hospital services. During the 1960s, dissatisfaction with the widespread patient movements into the cities led to a policy of building new district general hospitals in every district large enough to sustain one. By now, there are 15 general hospitals exceeding 300 beds outside the four cities. They are in Dumbarton, Greenock, Irvine, Kilmarnock, Dumfries, Kircaldy, Stirling, Falkirk, Inverness, Airdrie, Carluke, East Kilbride, West Lothian, Perth and Brechin (Institute of Health Service Administrators, 1984).

Matching hospitals to needs

The London Health Planning Consortium (1979) offers an example of planning the future distribution of hospital services. The consortium is made up of members from the four Thames Regional Health Authorities, the postgraduate hospitals, the Department of Health and Social Security and the University Grants Committee. It set itself the task of estimating how many hospital beds for each specialty will be required in each London health district by 1988. The objective was to identify a geographical distribution that is a fairer reflection of the needs of districts than the current pattern. A number of assumptions based on current trends in hospital care had to be made and, of course, the final result depends on their validity (Woods, 1982).

The consortium's method was to multiply detailed population projections by current hospital use rates (disaggregated by age, sex and specialty) to find the likely number of hospital patients in each specialty for every district in 1988. Trends in length of stay and turnover intervals in each specialty were projected to 1988 in order to calculate the number of beds required from the estimated number of patients in each district. Cross-boundary movements of patients were taken into account. The planning exercise allocated more hospital beds to inner-city areas than their population size and composition warranted because of two further adjustments. One was a modification to allow for a higher need for hospital services in certain districts due to unfavourable socio-economic circumstances. The other was to protect many of the beds of the inner-city teaching hospitals since medical students need access to sufficient numbers of patients with both common and rare conditions. In spite of both modifications, the overall effect was for the inner-city districts to lose hospital beds and for most districts outside the Greater London boundary to gain. Proposals based on a foundation of stated objectives and assumptions, as this is, are useful as targets towards which policy makers may move. Progress towards the target is inevitably hampered by more mundane considerations.

Planning and politics

Although planning the distribution of hospital services by mathematical formulae has an intellectual appeal, real decisions are taken with little room to manoeuvre between the constraints of limited finance, the availability of large plots of land suitable for development, the inertia of well established institutions and the interests of powerful groups of people both within and outside the NHS. Mohan's (1984a) study of the dispute about the location of a third district general hospital in Newcastle illustrates very well how far removed such decisions can become from considerations of patients' well-being.

The dispute extended from 1962 to 1969. The regional hospital board (predecessor of the health authority) had selected a suburban location in north-east Newcastle for the proposed new district general hospital because it was a virgin site, easy to develop and close to extensive housing. The proposal was vigorously opposed by several powerful groups which favoured a central city site. Consultants at the two existing district general hospitals in the city centre insisted that centralisation was essential for clinical reasons. Newcastle University authorities supported the central site because they wished to establish Newcastle as a major medical centre. The city council shared this aim, with the wider perspective of developing Newcastle as the regional capital. Other influential local figures saw a central hospital complex as an essential part of economic regeneration in north-east England. As the deadlock persisted, the regional hospital board became more entrenched in its position because not developing the available suburban site would have had a disruptive effect on the schedule of building elsewhere in the region.

Ultimately the Secretary of State (Richard Crossman) intervened, ostensibly as an independent arbiter, but in reality influenced by the pro-hospital-board briefing given him by his officials at the DHSS. Crossman resolved the dispute by supporting the hospital board's proposals in the main, but making one significant concession (on the location of the cardio-thoracic unit) to placate the most ferocious opponent, a thoracic surgeon. In this manner the hospital's location was fixed by the manoeuvrings of power blocs, each with its own objectives. Mohan observed that, throughout the dispute, virtually no attention was paid to the question of accessibility to the public or to the broader issue of whether the planned services were appropriate to the needs of Newcastle residents.

The location of hospitals has always been particularly subject to political influences. Hospitals are expensive facilities and they are vulnerable to variations in public expenditure. The plan to establish a network of district general hospitals is still not fully realised for that reason. Hospitals are also 'high profile' institutions which arouse strong reactions and might even win votes. Since the days of the municipal infirmary, decisions on the establishment, location or closure of hospitals have been swayed by the campaigns of local interest groups. Local politics continue to influence hospital planning, as indeed they should.

5

Transport to Health Services

Although there are important exceptions, most of the services
described in the previous two chapters are not usually supplied
in the consumer's home. Most consultations with family
doctors and dentists take place in surgeries and health centres.
General opticians and pharmacists normally work in retail
premises. These facilities are not often far from people's
homes, but hospitals are more widely spaced. A hospital
offering specialised services might be a considerable distance
from the homes of its patients. Whether near or far, most
health services require a journey and the nature of the journey
influences the use that is made of them and the benefits
obtained. Transport to health services is symbolised by the
ambulance, but the vast majority of trips to health services are
made independently. People who are ill or disabled are particularly likely to find a journey difficult. This chapter
describes the various transport links between people and
health services and seeks to identify the main problems associated with transport.

Modes of transport

The mode of transport used to get to health services varies
according to the type of service, the type of area and the
characteristics of the person concerned. In the OPCS study of
access to primary care (Ritchie *et al.*, 1981) 43 per cent of the
people who had visited the doctor in the past five years said
they usually walked to the surgery. A slightly lower proportion
travelled by car and only 15 per cent said they used public

Table 5.1: Modes of transport to doctor's surgery, by distance of surgery from home, in rural and non-rural areas (%)

Mode	Rural				Non-rural				Total
	< 1 ml	1–2 mls	2–5 mls	> 5 mls	< 1 ml	1–2 mls	2–5 mls	> 5 mls	
Walk	71	23	0	1	73	27	4	3	43
Public transport	1	14	16	10	4	26	43	23	15
Car	26	59	80	87	21	45	50	65	39
Other	2	4	3	2	2	3	2	6	3
Base	308	172	315	110	1,628	873	450	67	3,932

Source: Ritchie *et al.*, 1981, p. 24.

transport. There were large variations according to the distance to the surgery (Table 5.1). In both rural and non-rural areas, a large majority of people walked distances less than one mile, but few people walked more than two miles. The difference was made up by increases in public transport and car journeys, but public transport was much less used in the rural areas, and the car correspondingly more.

When asked how easy the journey was, 61 per cent of the sample replied that it was very easy, 33 per cent that it was fairly easy, 4 per cent that it was fairly difficult and only 1 per cent that it was very difficult. Different groups of people gave different answers, however. The people most likely to reply that their journey to the surgery was difficult were elderly, female, of social class V, travelling by public transport, living more than two miles from the surgery and registered with a large group practice. To illustrate, of the people living two to five miles from the surgery, 17 per cent of public transport users (all ages) reported some difficulty in the journey. At the same distance, 18 per cent of men over 65 and 26 per cent of women over 65 said they had difficulty.

Since a much smaller proportion of people live within walking distance of hospital than live close to a general practitioner's surgery, the distribution of modes of transport to hospital is quite different. Transport to four hospitals near Dudley and Stourbridge in the West Midlands was investigated by Andrew and Harris (1978). This is a largely urban area, with most trips from catchments of a ten-kilometre radius. Three of the hospitals were small general hospitals and the

Table 5.2: Modes of transport to West Midlands hospitals (%)

Mode	Staff	Visitors	Out-patients	In-patients
Car	54	75	51	42
Bus	33	19	26	4
Walk	11	4	7	< 1
Cycle	2	1	—	—
Ambulance	—	< 1	13	53
Other	< 1	< 1	3	< 1

Source: Andrew and Harris, 1978, p. 334.

fourth was mainly a long-stay unit. In all four, trips to work by staff exceeded the trips made by visitors, out-patients or in-patients, although in the hospital with a large out-patient department out-patients' trips approached the volume of staff travel. The travel modes of the various groups are shown in Table 5.2. Although 11 per cent of staff walked to work, the proportions of patients and visitors who walked were much lower. Similarly, staff were more likely to use public transport than patients and visitors, presumably for the same reason: most staff members lived close to their work. For out-patients and visitors the private car was the most important means of access to hospital. Most in-patients were transported by ambulance, but almost all the remainder were admitted after a car journey.

Whitelegg's (1982) survey of out-patients and visitors to eight hospitals in the Blackburn Health District gave very similar results. This district includes a large rural tract, but most of its population is urban. Even so, public transport was used by only a minority. Most out-patients and visitors travelled by car, even if this meant asking someone outside the household for a lift.

In a district where most of the hospital catchment area is rural, the private car is relied upon even more. Table 5.3 gives the mode of transport used for journeys to hospitals in an East Anglian health district with high car ownership levels (about 78 per cent of households had a car). Three-quarters of in-patients were admitted by hospital transport, mostly ambulances. Only one-tenth of out-patients were transported by the NHS. For out-patients and visitors the dominance of the private car is evident. Most car users travelled in their own car, but substantial proportions (8 per cent of in-patients and

Table 5.3: Modes of transport to hospitals in West Norfolk and Wisbech District (%)

Mode	In-patients	Out-patients	Visitors
Ambulance	72	2	0
Hospital car service	5	9	0
Private car	19	63	85
Walk or bicycle	2	14	10
Bus or train	1	10	3
Taxi	0	1	0
Base	250	722	316

Source: Haynes and Bentham, 1979b, pp. 89, 92, 95.

11 per cent of out-patients) relied on lifts given in someone else's car. All three categories of people were more likely to walk or cycle to hospital than to travel by public transport, although walking and cycling were used almost exclusively for trips of less than two miles. Public transport was used by few people to get to hospital, generally by people from non-car-owning households more than two miles from the hospital. The small proportion of people who made a trip to visit a patient in hospital by public transport is particularly notice-able.

PERSONAL MOBILITY

Constraints on mobility

Except for in-patients who may be transported by ambulance, access to health services beyond walking distance from home depends on the availability of a private car or a frequent bus service. Owning a household car is not the same as having it available, especially for people without a driving licence who rely on being driven, or for the people who are left behind at home while the household car is being used to transport somebody to work. Children who are too young to drive, elderly people and housewives may suffer low levels of personal mobility even in car-owning households. Possession of a second car removes some of these constraints. To com-pensate for long distances and infrequent public transport in

135

Table 5.4: Access to cars in relation to housing density for pensioners and young housewives (%)

Respondents	High density	Medium density	Low density	Rural
Pensioners:				
In car-owning household	18	30	58	30
In two-car household	2	5	12	9
With driving licence	2	7	29	16
Young housewives:				
In car-owning household	80	83	93	84
In two-car household	7	14	27	41
With driving licence	18	34	55	58

Source: Hillman and Whalley, 1975, p. 107.

rural areas, both car ownership and the proportion of people with a driving licence increase as housing density decreases. This was demonstrated by Hillman and Whalley's survey, whose results relating to two vulnerable groups in the population are given in Table 5.4. The proportions of pensioners and young housewives in car-owning households who possessed a driving licence themselves were lowest in high-density areas, where public transport is at its best. Although there was a general trend for car ownership and licence-holding to increase as housing density declined (and public transport worsened), the elderly living in rural areas had less apparent access to cars than their counterparts in low-density urban areas. The principal reason for this is the greater incidence of poverty in rural areas compared with the suburbs. In all four density categories, young housewives appeared to have markedly better access to cars than elderly people, especially in rural areas where 41 per cent were in households with a second car.

The main barriers to car use are low income, inability to drive and physical handicap. Each of these barriers affects the elderly more than any other group. Of people over 65, 76 per cent do not live in a car-owning household and 86 per cent do not have a driving licence (Norman, 1977). Physical handicaps may make it difficult to get in and out of a car, and they may also limit the distance it is possible to walk between the parked car at the end of the journey and the surgery or hospital. The Transport and Road Research Laboratory found that 31 per cent of their sample in a survey of elderly people living at home had difficulty getting around on foot (Richardson and Stroud, 1975).

Disabilities of this kind are more limiting for those dependent on public transport. Not only must passengers walk to and from the bus stops, they must negotiate the vehicle itself. Norman (1977) quotes a study made by British Leyland which estimated that about 4 million people in the country (2 million of them over 65) are too disabled to manage a step height of 17 inches, the legal maximum for bus platforms. Even heights of 10 inches would be too much for about half a million people.

A minority of the population, particularly amongst the elderly and people with low incomes, remains dependent on public transport. Within urban areas scheduled bus services provide a reasonably convenient means of access to doctor's surgeries, dentists, opticians and hospitals. In the most populated urban areas an hourly or more frequent bus service is available to virtually every household. By contrast, 10 per cent of households in rural areas do not have a bus service of even one per weekday, and only half the households in rural areas have an hourly or better service (Department of Transport, 1983). Rural public transport has dwindled considerably during the last twenty years and now consists mostly of services which run along the main roads connecting towns. In the West Norfolk and Wisbech District of East Anglia, for example, 12 per cent of the population were found to live in parishes with no bus service at all, and 42 per cent lived in parishes with less than 15 buses per day to any destination, which was considered the minimum level for hospital or surgery trips (Haynes and Bentham, 1979b). Figure 5.1 illustrates how bus services operate in corridors through rural areas, leaving the more isolated areas with little or no public means of contact.

For the majority of people without access to private cars in rural areas, the infrequency or absence of public transport is the main constraint on its use. Is there scope, then, for encouraging local bus transport to health services for people with no access to a car? Moyes' (1977) study, which simulated people's trips to the nearest surgery in Anglesey, suggests not. Moyes argued that potential surgery trips from households with no access to a car and away from the main rural bus routes are spatially dispersed and infrequent and therefore

Figure 5.1: Parishes with less than 15 buses per day, West Norfolk and Wisbech District

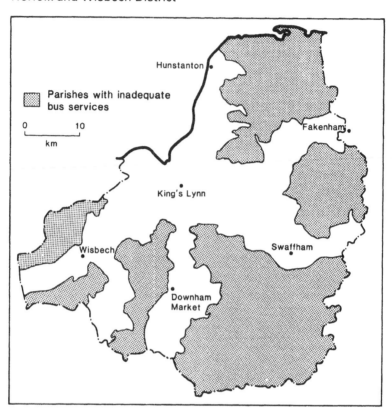

Source: Haynes and Bentham, 1979b, p. 30.

unlikely on their own to make the running even of a village minibus worth while. Marginal improvements in bus frequency in a scheduled rural service are likely to make little real difference to surgery accessibility. For the Anglesey population, Moyes concluded that reductions in the delay in getting to a surgery could be achieved more readily by increasing the opening hours of surgeries than by increasing the frequency of buses.

Andrew and Harris (1978) made the same observation in relation to transport to hospitals in the West Midlands. Even with only a ten-kilometre radius to cover, the demand for bus

138

transport to hospital was thought to be too small and geographically too dispersed to justify the provision of bus services.
Nevertheless, unconventional transport schemes in many parts
of the country have demonstrated that small-scale car and
minibus operations can be successfully used to make
services more accessible to the immobile (Moseley, 1979).

Transport costs

The most obvious costs of journeys to health services are those
of the time taken and the monetary cost of the trip. For most
car users, the costs of transport are likely to be trivial when
considered against the infrequency and high priority of health
care visits. The overall running cost of driving to the surgery
or to hospital is roughly twice that of the cost of the petrol
(Automobile Association Technical Services, 1977).

Both costs and time are more likely to be of concern to the
public transport user. Figure 5.2 illustrates information from
the National Travel Survey for 1978/9 and shows that about
half the people who travelled to their doctor's surgery by bus

Figure 5.2: Bus times to health services

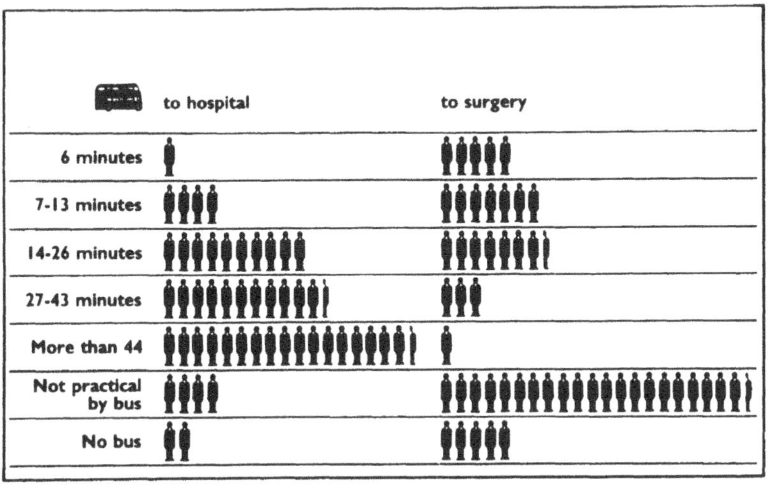

Source: Harrison and Gretton, 1984, p. 57.

had a journey of less than 13 minutes. As far as transport to the surgery is concerned, the availability of bus transport is a much more serious issue than the time it takes. Hospital trips, as might be expected, took longer, with the majority of journeys exceeding half an hour.

It is in the remoter rural areas that the monetary costs of transport reach significant levels. In South Cumbria, for example, a hospital appointment or a visit to a hospital patient typically involves return journeys of 60 to 180 miles, depending on the location of the hospital. Following a survey of people who had travelled at least 50 miles to get to hospital and back, Clayton (1984) observed the most serious difficulties applied to visitors rather than patients. Daily trips to visit hospital at the public transport rate, estimated in 1982 as 11p per mile, would cost a visitor £46 per week to a hospital in Lancaster and £138 per week to a hospital in Manchester. The trip to hospital also caused 18 per cent of the visitors in the sample to lose income. Clayton pointed out that the high costs of visiting in addition to the financial burden imposed by the loss of earnings from the patient can have severe consequences for some families. None of the visitors were eligible for ambulance transport, yet half of them said they suffered from a long-standing illness or disability. Public transport journeys to hospital were unpopular among both patients and visitors. The lack of co-ordination between transport and hospital times, long waits for transport connections, long walks to and from bus stops and the unpleasantness of long journeys while feeling ill were frequent complaints.

The discomfort, time and expense of the actual trip to hospital are not the only costs incurred. People not able or not confident enough to travel alone, working people who cannot easily take time off and young mothers with children, for example, must inconvenience other people as well as themselves. Table 5.5 summarises the incidence of these extra costs amongst out-patients questioned at hospital clinics in the West Norfolk and Wisbech District. It also indicates that, for some out-patients, the trip to the hospital clinic was combined with a trip to work, to school or to the shops: a fringe benefit which might be set against the overall cost of the trip. The fringe benefits reported rarely offset the inconveniences in individual cases, however. The people who combined their trip to the out-patient clinic with other activities tended to be younger

Table 5.5: Inconveniences and fringe benefits of the trip to an out-patient clinic, West Norfolk

Costs and benefits	% out-patients
Accompanied by another person	57
Lift in somebody else's car	11
Time off work (paid)	14
Time off work (unpaid)	9
Someone else off work	7
Arrangements for children	9
Combination with shopping trip	20
Combination with work trip	14
Combination with school trip	6
Combination with other purpose	5
Base	722

Source: Haynes and Bentham, 1979b, pp. 104–5.

adults from car-owning households relatively close to the hospital. These were not the people who reported the inconveniences.

Transport subsidies

Under certain conditions, financial assistance is given to patients with expensive journeys to hospital. In Scotland, there is a scheme administered by the Home and Health Department for residents of the Highlands and Islands. Patients who travel more than 30 miles to hospital or more than 5 miles by water may claim all but the first £1 of their travelling and unavoidable overnight expenses (Bain, 1983). There is a similar scheme for residents of the Isles of Scilly.

In England and Wales, the DHSS will pay the fares or petrol costs of in-patients or out-patients in three categories: war pensioners being treated for their pensioned disablement, people receiving family income supplement or supplementary benefit, and people whose family has less than £3,000 savings and whose fares leave them on a low weekly income. The third category of person is eligible on the principle that people with low incomes should not be worse off than those receiving supplementary benefit because of their hospital fares. Patients' escorts can receive payment for transport expenses if the hospital certifies that an escort is necessary. Visitors to hospital

141

patients can obtain help with their fares if they are visiting war pensioners or are on supplementary benefit and visiting a close relative. They must apply separately to a social security office. An explanation of the scheme and an application form containing detailed questions on income and outgoings are given in leaflet H11 *Fares to Hospital* (DHSS, 1984a), which is intended to be available for people to see in hospitals and doctors' surgeries.

Rosemary Lumb (1983) has evaluated the DHSS scheme from the point of view of residents of rural Northumberland, some of whom are far more disadvantaged in terms of time and cost than those on outer islands with air-strips. She found the scheme little publicised. The minority of doctors who did make the leaflet available in their surgeries nevertheless felt it was not their responsibility but that of the health visitor or social worker to make sure that the people who might have difficulty knew of the scheme. Other doctors said they advised people they thought might be eligible either to enquire at the hospital or telephone the DHSS. Northumberland hospitals, however, had the leaflet on display only rarely and usually could not answer enquiries about the scheme directly. One administrator suggested that making the leaflets more available would simply increase the number of ineligible people trying to claim. Lumb found that the onus was wholly on the patient to know of the scheme's existence and to persist with an enquiry. The process was likely to be discouraging.

Lumb also observed that the scheme is designed to help certain people who are not necessarily the ones most in need of assistance. After all, the ambulance service takes care of some hospital patients who have real transport difficulties. Problems with the cost of getting to hospital are more likely to be experienced by frequent visitors rather than patients attending for treatment, and not just by visitors who qualify for supplementary benefit or who are visiting a war pensioner. Even more significant in Lumb's view was the absence of assistance of any kind for travel to the services that are supposed to be local: to family doctors, dentists and opticians. These services are used much more frequently than hospitals but they also pose severe problems of access for some people.

An alternative scheme which meets some, but not all, of these points has been tried in Cumbria. The South Cumbria Travel Reimbursement Scheme was an experiment financed by the Development Commission (England's rural development agency) and administered by the South Cumbria Community Health Council to help some people living in that area with the cost of transport to hospitals. A typical journey for a patient in South Cumbria to and from a hospital in Lancaster is about 60 miles, although some patients must make round-trip journeys of 180 miles to Manchester. The purpose of the experiment, which ran for two years until late 1984, was to relieve the financial hardship of long journeys and to encourage attendance which might be inhibited by the journey. The beneficiaries were the patients of twelve selected general practitioners in rural Cumbria, who were reimbursed at half the public transport rate if they made a round trip of more than 50 miles (not using ambulance or hospital car transport) to attend a hospital as an in-patient or an out-patient. Similar payments were also made to people visiting either children or elderly patients in hospital. To apply for reimbursement, a form had to be filled in at the hospital and the money was later received by post from the community health council.

The scheme was evaluated by Susan Clayton (1984), who reported a number of difficulties which arose in publicising and administering it, several of which stemmed from the lack of enthusiasm for the scheme among the hospital staff whose active involvement was necessary. It was estimated that only about 20 per cent of patients who were eligible actually made a claim. The reasons given for not making claims were lack of knowledge or understanding of the scheme, the feeling that the sum of money involved was too small to be worth the trouble and a dislike of claiming assistance. Of the people who did claim, 26 per cent of patients and 6 per cent of visitors said that the journey to hospital had caused them financial hardship. These proportions did not vary significantly with distance to hospital, but they were related to income. Nevertheless, the scheme had been designed to make partial reimbursement available to all, irrespective of income, and the social class distribution of claimants was found to be similar to that of the local population. This might be interpreted as evidence of the scheme's wastefulness. The real difficulty is not simply distance (the people who live furthest away do not necessarily have the greatest problems), but car availability. Assistance might be more appropriately rationed according to income rather than distance.

VOLUNTARY TRANSPORT SCHEMES

In some localities there are voluntary transport schemes operated independently of the National Health Service which help to bridge the gap between immobile people and health facilities. Norman (1977) has described the wide variety of voluntary transport schemes available. The informal giving of lifts in a car belonging to another household is probably the most widespread practice, but large variations in the use of informal lifts have been noted from place to place, differences which probably depend on patterns of car ownership, employment, social structure and social cohesion. Lift-giving depends on one-to-one contacts and takes place between relatives, friends and acquaintances. Car pools are more organised schemes in which those who can offer lifts are put in touch with those who need them.

More complex schemes using voluntary transport in private cars are available here and there to meet special needs. There are several Women's Institute projects to provide transport for the elderly to get to hospital clinics, to visit hospital patients and to visit the chiropodist, for example. Some associations of hospital friends or other voluntary bodies organise transport for visitors to particular hospitals. Drivers in such schemes usually give their time for no payment, but receive mileage expenses which are financed by fund raising and by contributions from passengers. A few local authorities (at all levels from parish to county) have been persuaded to pay a mileage allowance and meet the administrative expenses of voluntary transport schemes organised by the WRVS or the Red Cross in rural areas. More ambitious projects use minibuses rather than private cars, vehicles usually acquired after fund-raising efforts and driven by volunteers. Voluntary minibus transport is commonly organised on an all-purpose, village or community-wide basis, but there is at least one health centre in Cambridgeshire which has used a minibus service, subsidised first by the Department of the Environment and subsequently by the county council.

Norman also outlines some of the problems involved. One perpetual difficulty is in getting and keeping enough drivers and organisers, and another is in financing the schemes. Local authorities subsidise public transport services but have in the past been reluctant to support competition which may ultimately kill the public bus.

Following the 1978 Transport Act, which relaxed the restrictions applying to community bus services and social car schemes, the DHSS encouraged health authorities to help promote social car schemes and also to do all they could to contribute to the success of community bus services in both rural and urban areas (DHSS, 1979c). Health authorities were asked to help community bus schemes by providing estimates of the demand for travel to hospitals and other NHS facilities, to provide parking space and garage facilities where possible, to loan NHS vehicles when the vehicles of voluntary organisations are temporarily out of action and even to contribute towards the cost of setting up a community service or the cost of running a service that is not self-financing. This advice, unfortunately, was given at a time when few health authorities felt they had resources to spare.

The hospital car service

The NHS hospital car service resembles some of the independent schemes described above, but it is financed by health authorities. Under this scheme non-emergency walking patients are transported to hospital by volunteers using their own cars and receiving mileage expenses. About 15 per cent of patients transported by the NHS travel by this means, covering about one-quarter of the total mileage (DHSS, 1982). Hospital visitors are not eligible for hospital car service transport. Not all health authorities operate a hospital car service. Of those that do, some are more committed to it than others. Depending as it does on the energy and enthusiasm of local voluntary organisers, the strength of the service is very variable geographically. Paradoxically, areas with high social needs and low levels of car ownership probably have the most difficulty in maintaining a car service. As Norman (1977) notes, the hospital car service could be used more efficiently (and would probably attract more volunteers) if hospital appointments were not bunched at the same time and if patients did not have to wait so long to be seen.

The success of the hospital car service depends not only on the supply of volunteer drivers but also on proper integration with the ambulance service. The advantages of using the two services in tandem have been described by the Chief Ambulance Officer of the former Devon Area Health Authority (Caple, 1976). In Devon, an integrated ambulance and hospital car service was operated through a central ambulance control with ambulance liaison officers at each of the four district general hospital complexes. Altogether 375 volunteer drivers were involved in the car service. In 1974/5 the car service accounted for approximately two-thirds of the total mileage and 61 per cent of the patients conveyed. As a result, Devon disposed of 25 per cent of its ambulance fleet as being unnecessary while absorbing an 18 per cent increase in overall workload. Caple claimed that use of volunteer drivers kept costs to a minimum and the reduction of the pressure of routine work improved the standards of the ambulance service for emergencies. The success of the scheme, in his view, was due to the co-ordinating role of the ambulance officers.

The flexibility of the hospital car service gives it an advantage over the conventional ambulance service for regular 'taxi' services, such as those necessary for day hospitals. In psychiatry, for example, day care is becoming increasingly preferred for people who would otherwise be long-stay in-patients and also for the elderly with organic brain disease. Howat and Kontny (1977) studied attendances at psychiatric day centres in Nottingham, where this type of treatment is well established.

In Nottingham, patients travelled to the day centres either by ambulance or by public transport (for which they were reimbursed by the hospital). The catchment area was compact (nowhere was more than eight miles from a day centre), but comparison of the average cost by public transport with the estimated cost by ambulance revealed that the ambulance was almost eleven times more expensive per attendance. The reasons given by hospital staff for allocating patients to ambulance transport were physical frailty (60 per cent), disorientation (17 per cent), lack of motivation to attend (9 per cent) and antisocial behaviour (7 per cent). Ambulance transport was considered essential for only 30 per cent of those allocated to it: the remaining 70 per cent were judged to be suitable for a car or taxi service if these became available. Working from the geographical distribution of the patients who could travel by car instead of by ambulance, Howat and

Kontny estimated the likely costs of a hospital car service which reimbursed drivers at standard rates and found that such an arrangment would cut the overall ambulance bill by 38 per cent.

Not only would it be cheaper, but other advantages could be expected from a mixed ambulance and voluntary car service. As it was, the ambulances used for psychiatric day patients had higher priorities (emergencies and transporting patients to out-patient clinics), so day-centre trips were made as and when possible, with the result that frequently up to half a patient's attendance time was lost. Howat and Kontny concluded that reducing the demands on an overstretched ambulance service could be expected to produce an increase in ambulance efficiency and possibly even allow a reduction in fleet size, as well as being beneficial for patients. They recommended the appointment of a part-time transport co-ordinator wherever psychiatric day centres are well established and argued that a voluntary car service could be backed up by using taxis where necessary, with substantial savings to the health authority.

There is some scope, then, for an expansion in the hospital car service if volunteers can be found and if the ambulance service can be persuaded to liaise. Bowen (1976) has suggested that more volunteers would be attracted if the service could be given a more local flavour rather than extending over a health district. He called for co-operation between hospital transport liaison officers and voluntary service organisers to establish hospitals' own voluntary car services, to be financed from ambulance funds like the existing car service. Such schemes have the blessing of the DHSS but would certainly encounter the opposition of the ambulance service if they were seen to be competing with the professional service for resources or patients. There is concern also over the safety of patients transported by the hospital car service, in the absence of controls to cover first aid training, driving competence and the condition of vehicles.

THE AMBULANCE SERVICE

Under the National Health Service Act 1977, the Secretary of State has a duty to provide sufficient ambulance services to meet all reasonable requirements. In practice, 'all reasonable

147

requirements' are those which fall into one of two categories: emergency transport to hospital and non-emergency transport which is deemed medically necessary. The current rules are given in a Health Service circular (DHSS, 1978a).

Anybody may request an ambulance for accidents or sudden illness in public places. Typically requests are made by a 999 telephone call. An ambulance must be despatched immediately to such calls: no medical authority is necessary. Emergencies account for about 5 per cent of all ambulance journeys (Levitt and Wall, 1984). For non-emergencies, the ambulance service provides transport to hospital or treatment centre for any patient who is considered by a doctor, dentist or midwife to be medically unfit to travel by any other means. Doctors, dentists and midwives are expected to take the decision about medical fitness themselves and not delegate it to receptionists, nurses, physiotherapists or anyone else. Ambulances are not intended to provide transport for patients visiting their general practitioner's surgery as general practitioners are required to make home visits to people who cannot get to the surgery because of their condition. Health centres are in an interesting intermediate position. Strictly speaking, an ambulance may not take a patient to a general practitioner's surgery in a health centre but may transport patients to specialised clinics in health centres.

The DHSS guidelines are clear that ambulance transport should be provided only for patients whose medical condition prevents them from travelling by other means. Being unable to afford transport does not qualify a patient for ambulance transport. It is acknowledged, however, that a more flexible interpretation of 'medical need' may be justified in rural areas where there is little public transport than in urban areas. A patient who could be expected to make a short journey to hospital might indeed suffer harmful effects after a long and tedious journey from a rural location. None the less, health authorities are instructed to ensure that demand for ambulances is limited to the essential.

There are 45 ambulance services in England. In London and metropolitan areas, regional health authorities administer the ambulance service. Elsewhere ambulance services are managed by district health authorities. Some district health authorities manage the ambulances for several districts, whose own authorities have no ambulance responsibilities. This is because the ambulance service had been managed by area health authorities until 1982, when area authorities were abolished. To disturb arrangements as little as possible, the area's role in ambulance management was usually taken over by a single district authority on behalf of all the area's districts.

There are three types of ambulance transport. Accident and emergency patients and those who are seriously ill or who cannot travel sitting up are assigned to the first category, which is a fully equipped accident vehicle with accommodation for two stretchers. These vehicles are used for routine clinic transport when they can be spared from accident and emergency duties. The second category is a sitting ambulance with two attendants, for patients needing assistance but able to travel sitting up. The third is a sitting ambulance with one attendant (the driver) for the patients who are least dependent. Although accidents and emergencies are only a small part of ambulance work, 90 per cent of crews are trained to deal with them. Most ambulance services switch crews on a shift basis between emergency work and routine out-patient work (National Association of Health Authorities, 1983).

The fully equipped accident and emergency vehicle is, of course, much more expensive to provide than the simpler ambulances used to transport sitting patients. This contributes towards higher costs in rural areas compared with urban areas since small ambulance stations serving rural areas must have the most specialised vehicle as a first requirement. Rural areas also have higher than average ambulance costs because they have a higher average mileage per patient than urban areas. Furthermore, rural patients are more likely to need ambulances than urban patients because the long journeys may be too much for them to manage.

Performance standards

The so-called 'ORCON' standards for ambulance performance were recommended to the DHSS by the Operational Research Consultants group at Cranfield Institute of Technology (Barnes et al., 1974). They were based on the standards which were actually being achieved by most of twelve ambulance services investigated. Different standards were recognised for ambulance services operating in metropolitan areas from those

Table 5.6: ORCON standards for ambulance services

| Measure of service | Percentile | Standard values | |
		Metropolitan service	Non-Metropolitan service
Emergencies:			
Activation time	95	3 minutes	3 minutes
Response time	50	7 minutes	8 minutes
	95	14 minutes	20 minutes
Urgent cases:			
Arrival time in relation to scheduled arrival time	95	5 minutes late	5 minutes late
Non-emergencies:			
Ambulance load to and from treatment centre	50	2 patients	2 patients
	95	8 patients	8 patients
Arrival time in relation to appointment time			
(a) Planned	5	40 minutes early	60 minutes early
	25	20 minutes early	30 minutes early
	75	15 minutes late	20 minutes late
	95	40 minutes late	60 minutes late
(b) Special	5	40 minutes early	60 minutes early
	25	20 minutes early	30 minutes early
	75	10 minutes late	15 minutes late
	95	20 minutes late	30 minutes late
Waiting time after treatment	50	30 minutes	30 minutes
	95	60 minutes	60 minutes

Source: Barnes *et al.*, 1974, p. 11.

of other ambulance services which generally had to cope with longer distances.

Table 5.6 summarises the standards of service which the Cranfield team found in their sample. Emergencies are the cases which require an immediate turn-out of an ambulance. The achievements of an emergency service are measured by activation time (the time between notification and the deployment of an ambulance) and response time (the time between notification and arrival at the scene of the incident). The table shows that the response time for 50 per cent of emergencies was 7 minutes or less in metropolitan areas and 8 minutes or less elsewhere. For 95 per cent of emergencies, response times were 14 minutes or less in metropolitan areas and 20 minutes or less in non-metropolitan areas.

Urgent cases differ from emergencies in that they do not require an ambulance immediately, but the patient must arrive at hospital within a specified time period (often one hour). For them, the critical measure of performance is the number of minutes late of the least efficient 5 per cent of cases transported. Efficiency in transporting the remaining non-emergency cases is measured in three ways. The first of these, ambulance load, is a surrogate measure for the additional journey time of the ambulance over and above the time it would take if the ambulance had only one passenger. Small loads mean less journey time for patients than large loads. Arrival time in relation to appointment time has such a wide variation that measurements were given for four different percentile points in the distribution. The investigators recommended closer liaison between hospitals and the ambulance service in order to reduce this range. Patients who were given special appointments were found to be delivered slightly more punctually. Finally, the waiting time after treatment was also considered to be capable of improvement with better liaison. These continue to be the critical measures against which ambulance services must measure their success (Steering Group on Health Services Information, 1983). In London, during 1984/5, only 45 per cent of emergency ambulance responses met the standards. The closure of a number of casualty units, heavier traffic in the capital and the severe winter were all thought to have contributed to the deterioration in service (Hencke, 1985).

Scheduling difficulties

Norman (1977) has outlined the difficulty facing the ambulance service, which is required to transport any patient for medical attention if a doctor says that ambulance transport is necessary. The service cannot therefore limit the demands made on it, yet it must work within a fixed budget. To make matters worse, hospitals cling to the habit of making all appointments for 9.00–10.30 a.m. or 2.00–3.00 p.m. which concentrates the demand for ambulances at these peak times. In rural Northumberland, Lumb (1983) found that 61 per cent

151

of out-patient appointments were for 9.00 a.m., an arrangement which produced enormous peaks and troughs in workload. From the patient's point of view the result is often three long waiting periods: for the ambulance, for attention at the hospital and then again for the ambulance. The journey itself is often roundabout and involves several stops, and is likely to be tiring and painful for frail passengers.

In her study of the transport needs of the elderly, Norman considered the shortfall in provision of transport to day centres to be particularly pressing, and found several examples over the country of day hospitals which were not being fully used because the ambulance service could not cope. In some cases, medical and ancillary staff were unoccupied for long periods because patients had not been brought in on time. Norman advocated attempts to increase the involvement of voluntary car schemes for transport to day centres, but warned that more than just transport is required. Often patients need help in finishing their dressing, turning off appliances and locking the house, and when they return they may need help again. This sort of sympathetic assistance together with the benefits of day centre therapy can make it possible for many elderly or handicapped people to continue to live at home.

Some of the problems are illustrated by Beer's study of mostly elderly patients attending Harrow Physical Treatment Centre (Beer *et al.*, 1974). Compared with other out-patient physiotherapy departments, the Harrow Centre had favourable transport arrangements. The catchment area is urban and compact, with most patients living within three miles of the centre. An ambulance and driver were reserved solely for the use of the centre, yet patients were found to spend an average of 2½ hours waiting for transport and travelling (and occasionally up to twice that time). The main source of delay was the block appointment system: all patients were told to be ready at 8.30 a.m. (if they were morning attenders) or 1.30 p.m. (for the afternoon session), but the ambulance would not usually arrive until much later. Early starts and frustrating waits were very disturbing to some patients. Non-walking patients could not use the regular ambulance and had to wait until a double-handed ambulance was free. For them, eight hours' waiting and travelling time were not exceptional.

Noting the disparity between waiting times and the half-hour spent in treatment, Beer and his colleagues were prompted to question whether the physiotherapy treatment was necessarily beneficial when weighed against the fatigue and anxiety generated by ambulance travel. There is clearly a threshold of waiting and travel which, if it is exceeded, outweighs the benefits of remedial treatment. Even with more efficient booking arrangements, the catchment areas for effective treatment in physiotherapy clinics and day centres are likely to be quite small.

A two-tier system?

The growth of day care for geriatric and psychiatric patients is one of the few expanding areas of health care, but it has placed great strain on already overburdened ambulance services. One way to relieve the pressure is possibly through a voluntary car transport scheme, as was suggested earlier. Another is to provide a lower-cost, second-tier ambulance service in addition to the emergency service. This has been tried at a geriatric day hospital in Lincoln which services a rural area with scattered population (Hyde, 1984). Previously up to six ambulances were used to transport patients to the hospital, four of them 'front-line', expensively equipped emergency vehicles. Patients commonly spent up to four hours in an ambulance for less than that time in 'day care'. They frequently arrived late and missed appointments with consultants. Sometimes, patients were still waiting at the hospital at seven in the evening.

Instead of continuing to use the normal ambulance service, Lincoln dedicated three vehicles (not equipped for emergency work) to serve the day hospital exclusively. Six ambulance staff were given reduced training and employed at a reduced salary to be the regular crew. A ten-mile catchment area was imposed and was divided into zones for more efficient routing. The zoning system meant that social services, home helps and meals on wheels services had to be reorganised for each patient so that he or she could fit in with the timetable. The new service is not saving money (it is provided in addition to the emergency ambulance service), but it has created real benefits. Journey times are much reduced and treatment times have been lengthened, so that the hospital now works much closer to its capacity. The new ambulance staff are regarded as part of the hospital team and are able to get to know patients, giving them personal attention and helping them with dressing and other domestic tasks, while keeping the hospital informed about home circumstances.

The success of this experiment lends support to the view that much of the routine transport work of the ambulance service might usefully be separated from the emergency function. Ambulance accident and emergency services might then be organised on similar lines to the other emergency services, the fire service and police, with a smaller supply of skilled personnel on call for real emergencies. A working party was established in 1980 to consider the organisation of ambulance services, but the subsequent Naylor Report rejected the idea of two separate ambulance services (Patient Transport Services Working Party, 1981). The report did accept that there is a case for experimenting with a two-tier system in some urban areas. In rural areas, the demand for accident and emergency transport was not considered to be consistent enough to permit an efficient service exclusively for that use.

There can be little doubt that using car transport or much simplified ambulance transport for the non-emergency ferrying of patients to and from hospital would be more economical than at present. It would also free ambulances and trained ambulance staff for accident and emergency work and the transport of severely disabled or seriously ill patients. In urbanised parts of Britain there may be enough work of this sort to sustain an exclusive specialised ambulance service. But the prospects of such a system being introduced are not great. As an editorial in the *British Medical Journal* (1977) expressed the problem, such a policy might be resisted both by the administrators, who would have to arrange and finance the new service, and by ambulance staff, who would fear that any change that lessened demands on their time would reduce their overtime payments and erode their industrial bargaining power.

Improvements to the service

Norman (1977) makes several suggestions for the improvement of hard-pressed ambulance services. Some are concerned with making sure patients receive the appropriate type of transport. She suggests, for example, that ambulance booking forms should be redesigned so that booking clerks do not automatically order a fully equipped accident vehicle when asked to order an 'ambulance'. Doctors should perhaps receive more training in the proper use of ambulance services, and physiotherapists and other professional staff could be given the authority to change the ambulance authorisation when a patient's condition has improved, without seeking a doctor's permission.

Other suggestions were incorporated into the DHSS guidelines for the use of ambulance services (DHSS, 1978a). Health authorities were reminded that out-patient and day hospital appointment times should pay due regard to the availability of public transport and the ambulance service. Block appointments are still common in out-patient departments, although the DHSS advice is that they should be avoided wherever possible, and that every effort should be made to phase out-patient appointments throughout the day to avoid peak loads on the ambulance service and reduce the anxious and often painful waiting periods for patients. The zoning of patients is also recommended to enable patients from the same locality to be collected and returned together. Such schemes require better communication between hospital staff and transport services than can be elicited by booking clerks. The DHSS accordingly advises health authorities to appoint ambulance liaison officers to help co-ordinate services, to deal with any problems relating to requests for patient transport and to ensure that there is no undue waiting or travelling for patients.

6

Health Care in Rural Areas

Social changes in rural areas

Within the last generation there have been considerable
changes in rural Britain. Most rural areas are increasing in
population, with younger families moving into owner-occupied
housing within commuting range of urban centres and elderly
migrants retiring to a house in the country, particularly in
coastal or scenic districts. This influx exceeds the gradual drift
away by rural residents, usually young people seeking employment and public housing opportunities, except in the remoter
areas, which are still losing population. The pattern of services, meanwhile, has been altering, with a trend towards
fewer, larger and more widely spaced facilities concentrated in
the more populous settlements. Small village primary schools
have been closed for reasons of both economy and quality of
education, but with little knowledge of the effects on young
children faced by longer journeys. Entertainment facilities
concentrate in the larger centres, as do shops. In villages with
less than 250 people, a single shop operates precariously.
Many village shop owners now approach retirement and their
businesses offer diminishing promise of profitability (Moseley,
1979). Health services, too, are concentrating. General practitioners' surgeries are disappearing in the smaller villages as
doctors have joined together in group practice and many small
local hospitals have been closed, to be replaced by a district
general hospital in a town some distance away.

For most rural inhabitants, the centralisation of services
has not made them notably less accessible because it has
been accompanied and encouraged by marked improvements

156

in personal mobility. It is not uncommon for three-quarters of the households in rural areas to own at least one car and the proportion is still rising. At the same time, however, the availability and use of public transport in rural areas are in decline, locked in a vicious circle of lower use leading to higher fares and lower frequencies, which in turn make the rural buses even less attractive to use. Regular rural bus services have now largely disappeared. What is left is an inter-urban service which benefits only rural people who happen to live near the main roads between towns. A minority of people, the ones without access to a private car, have been left with worse access to basic services than in former generations. This minority includes those who are too poor to afford a car, the elderly and disabled who cannot drive, and, during the day, women and children who are without transport while the household car is used as transport to work. All these groups are expected to have a higher than average need for health services.

Health in rural areas

The view that people in rural areas live long and healthy lives compared with city dwellers is widely held. It is supported by evidence on mortality which shows that, when standardised for age and sex, death rates in rural areas are lower than those in urban areas for every major cause of death except road accidents. Closer examination of mortality rates, however, reveals that the generally low figures for rural districts of England and Wales obscure marked differences between relatively high rates in the remoter rural areas and lower rates in the growing areas close to the main towns. Mortality rates are negatively associated with population change, with the highest mortality being found in the areas sufering population decline (Bentham, 1984). It seems likely that the people leaving the declining rural areas have been disproportionately the young, better qualified and healthy, and that the least healthy have tended to remain in the remoter areas: the same areas which are experiencing the greatest loss of services.

Mortality rates are, at best, a very crude indicator of general health (and a poor measure of the demands on health services), so it is of interest to know whether the pattern of

157

mortality is matched by the distribution of morbidity. Morbidity is a notoriously difficult variable to measure without introducing bias, but perhaps the best national source of information is the General Household Survey carried out annually by the Office of Population Censuses and Surveys. The General Household Survey results are not published for local areas, but the same morbidity questions were used in a survey of Norfolk by Bentham and Haynes (1985). This showed that villages in remoter areas (15–21 miles from the city of Norwich) had higher rates of limiting long-standing illness and short-term illness than villages close to the city. The highest rates of illness were found in the remoter villages which had no general practitioner's surgery. Of the people aged 65 years and over, for example, 28 per cent reported acute illness during a two-week period in the remote villages without a general practitioner's surgery, compared with only 14 per cent in villages with a surgery and close to the city. These differences in morbidity were compounded by the presence of greater than average proportions of elderly people in the least accessible villages.

The relationship between health and personal mobility adds another dimension to the situation. People with low levels of personal mobility are generally those such as the elderly and the poor who are most at risk from ill health. Table 6.1 illustrates the situation in the city of Norwich and a selection of Norfolk villages, close to Norwich and more remote, some of which have general practitioner surgeries and some not. Two trends are strong. Households which do not have cars had higher morbidity levels than car-owning households, particularly for chronic illness. Furthermore, households with similar mobility characteristics were more likely to suffer illness in the remoter locations. Comparing households with a car and telephone in the city with households without car or telephone in the remoter villages which have no general practitioner surgery reveals a threefold difference in short-term illness and almost a fourfold difference in long-term illness. Lack of easy access to a car or a telephone is, of course, a significant barrier to communication with general practitioner surgeries which use appointment systems and are located several miles away. People in remote rural areas appear to be doubly disadvantaged: firstly by their poorer levels of health and secondly by their difficulties of access to health services.

Table 6.1: Self-reported morbidity, personal mobility and accessibility to health services in Norfolk

	Norwich	Near Norwich		Far Norwich	
		With GP	Without GP	With GP	Without GP
	% of respondents ill during previous 2 weeks				
Household has car and telephone	7.3	11.4	11.4	10.7	13.5
Household lacks car	13.0	13.5	12.2	14.1	20.3
Household lacks car and telephone	16.3	6.3	11.1[a]	10.5	21.3
	% of respondents with limiting long-standing illness				
Household has car and telephone	8.6	9.0	7.8	11.6	12.7
Household lacks car	23.2	21.6	31.0	27.7	29.3
Household lacks car and telephone	30.2	9.4	21.1[a]	30.8	33.3
Base	277	453	279	316	278

Note: a. Based on a sample of less than 30.

Source: Bentham and Haynes, 1985, p. 235.

EFFECTS OF INACCESSIBILITY

The most obvious disadvantage of living at a greater than average distance from health services is that the costs of receiving them are relatively high. The inhabitants of rural areas pay more for the benefits of the National Health Service than city and suburban residents because they spend more on petrol and occasionally on bus fares, they lose more time travelling and they must cope with an increase in the non-monetary inconveniences that longer journeys entail. Usually these are not serious problems. For most people, including those who must rely on public transport, a trip to the doctor, dentist or hospital is of sufficient priority to make considerations of cost, time or inconvenience relatively trivial. There are two main exceptions. In the most isolated rural areas, where return journeys to hospital could exceed 200 miles (Bain, 1983) and may involve an overnight stay, the cost of a single trip are not inconsiderable. Although financial assistance is available for those who claim it, reimbursement does not cover loss of earnings or any other incidental expense. The second exception is the case of people who must make regular journeys to hospital. This includes people who need repeated out-patient treatment (such as radiotherapy) and also regular visitors to hospital patients, for whom no transport subsidy is provided (unless they are visiting war pensioners or receiving supplementary benefit). Even quite modest journeys can become a significant drain on money, time and energy when they must be repeated regularly over an extended period. Rural residents are at a serious disadvantage in this respect.

A second disadvantage of living a relatively long distance from health services is potentially much more damaging. In some cases the barrier of inaccessibility may inhibit the use of services (Joseph and Phillips, 1984). If there is evidence that the use of primary care or hospital services is sensitive to distance, then it will be apparent that rural areas remote from services are receiving less than their share of the benefits of the National Health Service.

Accessibility and general practitioner consultations

The hypothesis that physical accessibility may significantly affect the use of health services in Britain goes against a widespread assumption in the National Health Service that travel to services is not a serious problem except in the remoter parts of Scotland. There is some evidence to support this view. A Royal Commission study (1979b) found that travel distances in a mixed urban and rural area of Cumbria did not seriously limit access to primary care except for a minority of elderly people who were deterred by the physical effort involved in the journey. A similar survey by the Welsh Consumer Council (1979) also failed to reveal any widespread problems of locational disadvantage. Both studies, however, were more concerned with attitudes towards access than objective measures of deterrence. Attitudes and expectations, it is well known, mould themselves to prevailing circumstances. Neither study compared consultation rates of people living close to services with those of similar people living further away.

Geographical variations in consultation frequency cannot be detected without first taking into account the differences which

160

Table 6.2: Number of consultations at surgery during previous year, by distance from surgery (%)

Informant consulted GP	< 1 ml	1–2 mls	2–5 mls	> 5 mls
Not at all	27	26	32	34
Once only	18	21	18	20
2–3 times	25	24	24	21
4–5 times	11	9	11	8
6–10 times	9	11	8	10
More than 10 times	10	10	7	6
Base	2,102	1,124	842	202

Source: Ritchie *et al.*, 1981, p. 47.

are expected on the basis of age, sex and social class. Elderly people consult more frequently than the young, women more than men, and the unskilled more than professional groups. Health itself is, of course, the other important influence. People who assess their own health as poor consult much more frequently than those who say their health is good, although in one national survey 5 per cent of those saying their health had been poor over the preceding year had not consulted at all during that time (Ritchie *et al.*, 1981). Health is also the most difficult variable to control and most studies have not attempted the task.

In their national sample of 4,733 respondents, Ritchie *et al.* (1981) found that distance to the surgery had a slight deterrent effect on consultations with a general practitioner (Table 6.2). While 27 per cent of people living less than one mile away from their surgery had not consulted at all during the year, the proportion was slightly increased to 34 per cent at five miles or more. Conversely, the proportion of people consulting more than ten times per year was slightly reduced by increasing distance. These results, however, were not controlled for any of the variables which would tend to obscure the effects of distance.

Parkin (1979) was able to control for demographic and social variables and demonstrate that general practitioner consultation rates are sensitive to small distance variations in an urban area where most people live close to the surgery. He analysed the records of a single general practice in the London borough of Lambeth and examined the annual consultation rate for three distance zones, with boundaries at 0.4 kilometres and

Table 6.3: General practitioner consultation rates in an urban practice by distance zones

Population group	Consultation rate			
	Overall	< 0.4 km	0.4–1 km	> 1 km
Whole population	4.82	5.07	5.18	3.53[a]
Males	4.17	4.17	4.80	3.37[a]
Females	5.40	5.89	5.53	3.67[a]
Social classes I and II	4.14	4.29	4.98	2.45[a]
Social classes III, IV and V	4.88	5.14	5.21	3.63[a]
Age groups (years):				
0–14	4.37	4.55	4.87	3.13[a]
0–14 (male)	4.18	4.16	5.17	2.88[a]
0–14 (female)	4.55	4.90	4.59	3.35[a]
15–64	4.49	4.55	4.94	3.69[a]
15–64 (male)	3.73	3.57	4.33	3.46
15–64 (female)	5.22	5.52	5.52	3.91[a]
Over 64	7.33	8.67	6.96	3.34[a]
Over 64 (male)	7.08	8.20	6.75	3.86[a]
Over 64 (female)	7.49	8.97	7.12	3.02[a]

Note: a. $p < 0.01$ for difference with zones 1 and 2.

Source: Parkin, 1979, p. 97.

1 kilometre from the surgery. Generally the differences in consultation rates between the first two distance zones were small. However, consultation rates fell for people living more than one kilometre from the surgery. The reductions were statistically significant in every group except for males aged 15–64 (Table 6.3). One possible explanation for males of working age not conforming to the general trend is that this group is more likely than any other to visit the surgery from work rather than from home.

Results such as these must be interpreted with care. While it is usually assumed that higher consultation rates indicate higher-quality contacts between patients and doctors, this is not necessarily the case. Doctors and patients who communicate well may establish lower than average consultation rates by reducing the number of unnecessary consultations. In general, however, a consultation rate that is lower than expected is taken to indicate some hindrance to access. There is also the consideration that low consultation rates at the surgery might be compensated by increased home visiting by the doctor, although little or no evidence is available to confirm this hypothesis. Ritchie and her colleagues (1981) found that people in rural areas did not receive more home visits than people in non-rural areas. For the practice studied by Parkin, home visits were actually a higher proportion of all consultations within one kilometre of the surgery than outside that radius (Morrell *et al.*, 1970).

Use of out-patient services

Distance from the general practitioner surgery may have some effect on a patient's likelihood of being referred to an out-patient clinic. In her study of out-patient referrals in the Scottish Borders Region, Rosamond Gruer (1972) compared the referrals of patients living less than three miles away from their doctor's surgery with those of patients more than three miles away. The rates were 10.0 referrals per 100 population per year for the people living less than three miles away and 5.6 referrals per 100 population per year for the more distant population, a difference which is statistically significant at the 0.001 level. The fall in referral rates with distance from the surgery was found to apply to most of the common diagnoses. The exceptions, where there was no apparent reduction in referral rates with distance (or only a small one) included the most life-threatening conditions of malignant and heart disease. Some of the conditions whose referral rates seemed to be dramatically associated with distance were the less serious ones such as varicose veins, haemorrhoids, hernias and minor ear, nose and throat conditions, which may not worry many individuals sufficiently to overcome the difficulty of access to their general practitioner. The trend, however, was strongly apparent across many diagnostic groups.

Gruer's study also raised the possibility that referral to an out-patient clinic depended not only on access to the general practitioner surgery but also on the distance to the clinic itself. In the four Scottish Border counties at the time of her study, out-patient referral rates per 100 population per year (standardised for age and sex) varied from 5.7 (Berwick) through 6.4 and 8.3 (Peebles and Roxburgh) to 9.2 (Selkirk). Gruer noted that in terms of a crude index of distance to hospitals, the counties ranked in the same order, with Selkirk the most accessible and Berwick the most remote. She also drew attention to the difference between the crude referral rate in the

Table 6.4: Actual and expected distribution of out-patients by distance from King's Lynn

Distance	Actual out-patients	Expected out-patients	Ratio of actual to expected
Within King's Lynn	138	83	1.7
10 miles or less	79	74	1.1
Over 10 miles	90	150	0.6
Base	307	307	1.0

Source: Haynes and Bentham, 1979a, p. 190.

rural Borders area (8.7 per 100 population) and that of Edinburgh (15.2 per 100 population), where hospital services are much more accessible. Circumstantial evidence of this type which is not corrected for any other possible influence must, of course, be interpreted with some scepticism.

In a study in rural East Anglia, Haynes and Bentham (1979a) defined three concentric distance zones around the town of King's Lynn, Norfolk, and interviewed 307 people attending out-patient clinics in the town to find out in which zone they lived. The actual numbers of people in each zone were compared with the numbers expected if distance had no effect. Expected numbers were calculated by weighting the age/sex composition of the population of each distance band by the corresponding national age/sex-specific out-patient attendance rates. Table 6.4 gives the results and shows a marked decline in the ratio of actual to expected out-patients with distance from the clinic. The samples were small and not strictly random, and the results may also be affected by some patients being referred across the health district boundary to a more distant out-patient clinic (20–40 miles away). Nevertheless, the apparent influence of distance is strong.

There are several possible reasons for a decreasing use of out-patient clinics with increasing distance. It has already been shown that people living in the less accessible rural areas are unlikely to be healthier (the reverse is more probable), but perhaps they have lower expectations of health care. Perhaps their general practitioners are more self-reliant than general practitioners working closer to the hospital and consequently less likely to refer patients to hospital. Another possibility is that the difficulties of travel discourage people from agreeing to an appointment or even keeping an appointment in a hospital clinic some distance away. Gruer's suggestion that distance to the general practitioner surgery also affects out-patient referrals complicates the situation. Is there, perhaps, a link between reduced consultation with family doctors in more remote rural areas and reduced out-patient referrals? Can any distance effect be detected in in-patient admissions? These hypotheses ideally require longitudinal studies which trace the whole process from illness to consultation, referral and then hospital admission.

Morbidity, personal mobility and use

Haynes and Bentham's (1982) survey in Norfolk was designed to examine any effects of accessibility in a rural area on general practitioner consultations, out-patient attendances and in-patient admissions. It was a cross-sectional investigation at a particular time rather than a more rigorous (and difficult) longitudinal approach tracing people's experiences over a period of time. Nevertheless, it offered the possibility of discerning links in the progression from primary to secondary services and it attempted to control for variations in both illness and personal mobility, two of the most important influences on service use. Random samples of the population were interviewed in the city of Norwich and in 16 villages served by the two general hospitals in Norwich (and no other general hospital). The villages were selected in equal numbers from four categories: (1) near to Norwich (4–7 miles) with a general practitioner surgery in the village, (2) near to Norwich but with no surgery in the village, (3) far from Norwich (15–21 miles) with a surgery, (4) far from Norwich with no surgery. This design enabled the effects of remoteness from the surgery and remoteness from the hospital to be separated. Altogether, 1,603 people were questioned on their personal mobility, health and use of health services.

Measures of both chronic and acute morbidity were used. Information was collected on long-standing illness that limited the respondent's activities and on short-term restricted activity during the two weeks prior to the interview. The use of health services was highly associated with both measures, as might be expected. To cancel this effect and control for geographical variations in illness, 'use/needs' indices were calculated

Table 6.5: General practitioner consultations: use/need ratios

| | | Use/need ratio | | | |
| | | Near Norwich | | Far Norwich | |
	Norwich	With GP	Without GP	With GP	Without GP
Household has car and telephone	1.81	1.40	1.52	1.09	0.64
Household lacks car	1.23	1.01	0.68	0.85	0.69
Household lacks car and telephone	1.06	1.40	0.71[a]	1.08	0.57

Note: a. Based on a sample of less than 30.

Source: Bentham and Haynes, 1985, p. 236.

(Bentham and Haynes, 1985). For general practitioner services, the index was the percentage of respondents consulting a general pratitioner in a four-week reference period divided by the percentage of respondents with limiting long-standing illness or short-term restricted activity. High numbers indicate high levels of use relative to needs, and vice versa.

The ratio of general practitioner consultations to illness needs is given in Table 6.5 for people with different levels of personal mobility and in areas of different accessibility. Two trends are clear. Ownership of a car and a telephone appears to increase consultation rates relative to need, and consultation rates increase with increasing accessibility to surgeries and the city. Households in the city with a car and a telephone have consultation rates three times higher relative to their needs than households remote from the city with no surgery in the village and with no household car or telephone.

Use of hospital services was assessed by measuring the percentage of respondents who had been a hospital in-patient or out-patient or who had attended an accident or emergency department in the previous year. Again, these percentages were divided by the same measure of morbidity, to produce use/need ratios. In Table 6.6 the hospital use/need ratios are disaggregated by personal mobility and accessibility. This table shows that the hospital services of Norwich tend to be used disproportionately by the more mobile people living close to them. Households in the most remote villages, particularly those with low personal mobility, have very low levels of use relative to their health needs. For casualty and in-patient

Table 6.6: Hospital services: use/need ratios

	Norwich	Near Norwich		Far Norwich	
		With GP	Without GP	With GP	Without GP
Out-patient:					
Household has car and					
telephone	1.78	1.20	1.06	1.07	0.52
Household lacks car	1.12	0.82	0.83	0.66	0.53
Household lacks car and					
telephone	0.81	1.21	0.95[a]	0.71	0.43
Casualty:					
Household has car and					
telephone	1.21	0.49	0.71	0.49	0.16
Household lacks car	0.39	0.05	0.00	0.25	0.22
Household lacks car and					
telephone	0.19	0.32	0.45[a]	0.31	0.18
In-patient:					
Household has car and					
telephone	0.55	0.45	0.71	0.52	0.29
Household lacks car	0.19	0.32	0.45	0.31	0.18
Household lacks car and					
telephone	0.19	0.60	0.00[a]	0.32	0.05

Note: a. Based on a sample of less than 30.

Source: Bentham and Haynes, 1985, p. 237.

services there is a greater than tenfold difference between the two extreme groups.

Comparing the use of health services of the people who reported a long-standing illness with the people not suffering from long-standing illness separated two components of the accessibility effect. One component was a very high demand for health services by healthy, mobile and prosperous people with high expectations and within easy reach of general practitioner and hospital services. The other was a reduced demand in rural areas by non-car-owning people with long-standing illness. People who reported long-standing illness which limited their activities (more serious chronic illness) showed the largest reduction in demand: an indication that the problem affects those with the greatest need.

It seems from Tables 6.5 and 6.6 that most of the effects of accessibility on use are present at the initial stage of general practitioner consultation, where the availability or remoteness of a surgery is the key factor. Even in the measures of hospital use, the distinction in the villages more distant from Norwich between higher usage rates in the villages containing a surgery and lower rates in those with no surgery are preserved. A separate, but perhaps smaller, effect of hospital inaccessibility is also suggested.

Branch surgeries

The tendency for branch surgeries to close down in rural districts away from the main towns (Chapter 3) is a regressive trend in terms of access. Branch surgeries have limited opening hours, they are often poorly equipped and offer few comforts, yet they have long been useful points of contact between the doctor and less mobile patients who live in outlying settlements.

Fearn (1983) has surveyed the branch surgeries in the rural parts of Norfolk. He found that the majority were held in private houses or village halls, with about one-fifth in purpose-built premises, which ranged from permanent buildings to small wooden huts. A few occupied exotic premises such as public houses, disused barns and a railway station. Opening hours varied considerably, but most branches were open one or two days a week for one to three hours. The majority, but not all, had a waiting area for patients, a consulting area with adequate privacy, heating, a toilet and hot water. Relatively few had a receptionist or the clinical equipment that would be expected in a main surgery. Less than half the branch surgeries had patients' clinical notes on the premises. In general, the purpose-built premises were the best equipped, open for the longest hours and with increases in attendance in the last five years. Branch surgeries held in private houses and village halls had fewer facilities, fewer patients using them, and often showed a decline in attendance in recent years.

These differences were also reflected in the characteristics of patients attending. While doctors reported that they saw similar types of people in purpose-built branch surgeries as in their main surgeries, doctors attending village halls and private houses said they saw greater proportions of elderly patients, females, working-class people, people without cars and patients with less serious conditions than they did at the main

Table 6.7: Users of branch surgeries by age, sex, occupational group and car ownership

	Number of patients	Percentage who use a branch surgery			
		Always	Mostly	Sometimes	Never
Age (years):					
18–44	79	9	5	35	51
45–64	78	9	23	33	35
65+	74	35	23	11	31
Sex:					
Male	105	14	19	24	43
Female	125	19	15	30	36
Social class:					
Non-manual	77	8	4	31	57
Manual	142	21	25	26	27
Car ownership:					
Car owner	166	7	13	31	50
Not car owner	64	44	27	17	12

Source: Fearn et al., 1984, p. 490.

surgery. A separate survey of residents in villages with branch surgeries of the less well appointed variety confirmed most of these impressions. There were clear gradients in frequency of use associated with age, occupation and car ownership, although not with sex (Table 6.7). Precisely the same groups commented unfavourably about the time and cost involved in getting to the main surgery.

Neither the doctors nor the patients who responded to Fearn's survey had any illusions about the generally poor facilities and primitive conditions of branch surgeries. The doctors felt that these implied low levels of care, but the real difficulties of access to the main surgery for people unable to drive or with no household car must be balanced against this. For these groups, the presence of a branch surgery in the village appeared to increase consultation rates substantially. Table 6.8 compares consultation rates in different types of village. Consultations at main surgeries dropped by one-quarter in villages lacking a main surgery compared with villages with a main surgery, but branch surgeries more than compensated for this. The result was that villages with branch

Table 6.8: Percentage of respondents who consulted a general practitioner during previous four weeks: the effect of branch surgeries

| | Place of consultation | | | |
	Main surgery	Branch surgery	Home	Telephone
Villages with a main surgery (N = 316)	17.4	0	2.8	1.3
Villages with a branch surgery (N = 282)	12.8	7.8	2.8	0.4
Villages with no surgery (N = 342)	12.9	0	2.3	1.8

Source: Fearn *et al.*, 1984, p. 490.

surgeries had markedly higher consultation rates than villages with no surgeries. In the villages with no surgery facilities, telephone consultations were slightly increased and home visits by the doctor were slightly fewer. These other forms of consultation did not therefore compensate for the lack of a surgery.

Hospital visiting

Visitors to hospital patients might also be deterred by the difficulties of getting to hospital from rural areas. If so, this might be expected to affect the morale and perhaps even the medical condition of 'neglected' hospital patients. As well as having a therapeutic value, visiting is beneficial in that it maintains the links between patients and their home communities (discharge is difficult for patients and their families if these links are allowed to lapse) and, furthermore, visiting gives hospital staff the opportunity to discuss the patient's condition and arrangements for his or her care after discharge. Hospital visiting hours have become increasingly flexible in recent years, especially in long-stay hospitals, in many of which visiting is permitted at any time during the day.

At least three basic measures of hospital visiting could be used. The number of visits refers to the total number of people received by patients over a period of time: a number which may include particular visitors more than once. The number of

Table 6.9: Respondents in accessible and inaccessible villages who had visited hospital patients in previous year (%)

Characteristics of respondent	Accessible	Inaccessible
Age:		
16–44 years	48	49
45–64 years	41	38
65 years and over	47	29
Sex:		
Males	38	37
Females	50	40
Car ownership:		
Car in household	47	41
No car	37	27
Total	46	39
Base	268	243

Source: Haynes and Bentham, 1979b, p. 122.

visitors, on the other hand, is the total number of different people visiting patients. Researchers more interested in transport may also measure the number of visitor trips. This is the total number of groups of people who travelled together to the hospital in a time period, a number which, like the number of visitors, could include particular people more than once.

It can be shown that residents of rural areas from which hospitals are relatively accessible are more likely to visit hospital patients than residents of villages more remote from hospitals. Table 6.9 is based on a survey of six accessible villages (4–7 miles from hospital) and six comparatively inaccessible villages (14–19 miles from hospital) in West Norfolk. There were marked differences in the proportion of respondents who had visited during the previous year in the two types of village for particular population groups. The elderly, women and people in households which did not have a car were less likely to have visited a patient from the inaccessible villages than from the accessible villages. These groups in the remoter villages were aware of their disadvantage and expressed much higher levels of dissatisfaction over the difficulty of getting to hospital than all other groups.

Two effects of inaccessibility on visiting rates may be investigated: the relationship with distance from the visitor's origin to the hospital and the effect of distance from the patient's home to the hospital. These coincide only for the visitors whose trip origin is the patient's own home. Of the two, the distance or difficulty of the actual visiting trip is more fundamental as an explanation of any diminution in visiting. Nevertheless, since it is much easier to establish the geographical origins of patients than those of visitors, the question whether patients who live close to the hospital have more visitors than those who do not is of more interest from the planning point of view.

Any investigation of visiting frequencies to particular patients must take into account the duration of hospital stay. Visiting declines as the duration of stay of the patient increases. Cross and Turner (1974), for example, found that for geriatric patients increasing the hospital stay from one month to six months had the effect of halving both the numbers of visits and the number of visitors.

Cross and Turner's study was of visiting patterns to patients in geriatric units in Shropshire. Their results indicated that short-stay geriatric patients were not likely to receive fewer visits if they lived far from the hospital compared with those who lived closer. For long-stay patients, however, there was a marked diminution in visiting with increasing distance from their homes to the hospital (Table 6.10). Cross and Turner concluded that 'assessment' units in which geriatric patients are kept for only short stays can be sited within any distance (up to 32 kilometres) of the homes of patients without adversely affecting visiting rates. They recommended that patients needing custodial care over a long term should be accommodated in hospital units within 16 kilometres of their own homes, because distance probably does affect visiting rates for this category of patient.

In the West Norfolk study (Haynes and Bentham, 1979b) two-thirds of the hospital visitors surveyed lived within two miles of the patient they were visiting. Visitors to preconvalescent patients (a sample of patients in acute medical and surgical beds who had completed the most intensive part of their treatment), geriatric patients and elderly mentally infirm patients were counted over a period. The elderly mentally infirm patients in the sample were visited so infrequently (an average of two visitor trips per month by a single visitor) that no trends were distinguishable. Visiting to geriatric patients was more frequent, with averages of three visitors and three visitor trips per week. As Cross and Turner had found, length

Table 6.10: Percentages of geriatric patients who received more than two visits during the week

Distance of patient's home from hospital	Length of stay of patient			
	< 2 months	2−6 months	6−36 months	> 36 months
0−8 km	88	72	53	39
9−16 km	71	71	58	39
17−24 km	86	81	46	22
25 + km	88	71	24	17

Source: Cross and Turner, 1974, p. 137.

of stay significantly affected visiting to geriatric patients. A possible weak relationship with distance was not statistically significant. For preconvalescent patients there was a significant association between visiting and both duration of stay and distance from home. These patients received 14 visitor trips (comprising six visitors) on average per week, but the numbers varied with length of stay and home distance.

Although it is difficult to demonstrate a distance effect for patients who are visited very infrequently, younger, acute patients living close to the hospital appear to receive more visits than similar patients from further away. Higher visiting rates are therefore expected to be a substantial benefit of smaller, more local hospitals compared with larger district units.

Accident and emergency services

For the accident and emergency service, accessibility might literally be a matter of life or death. As was described in Chapter 4, the trend over the last two decades has been to concentrate accident and emergency services into large units serving a minimum catchment of 150,000 population. Medical opinion is that serious accident and emergency patients are best treated in well equipped units by specialised staff who are available at all times. This almost unanimous view is based on experience of the benefits of larger units. For the most serious accidents and emergencies, lives might be saved in a special-ised unit that would be lost in a small local facility. Much less is known about the costs of longer journeys in terms of their effects on patients whose condition is deteriorating rapidly. Here, also, lives may be at stake.

Some information on this issue is supplied in a study of deaths due to motor vehicle accidents in males aged 15–24, the group with the highest traffic accident death rate. Bentham (1986) examined the variations in this death rate between 401 English and Welsh county districts, metropolitan districts and London boroughs, and found that death rates in the most rural areas were more than double those of large towns and city areas. Multiple regression analysis revealed that car ownership levels (a surrogate for variations in the use of private transport) and the percentage of people in social classes IV and V were both positively and significantly correlated with traffic accident death rates at the local authority level. These factors in combination could be partly responsible for the predominantly rural nature of traffic deaths. People in rural areas are much more dependent on motor vehicle transport than urban residents and are therefore exposed to more risk. These are also possible explanations for the relationship with social class, connected with motor cycle riding, alcohol drinking and, perhaps, attitudes towards risk taking. These were not the only relationships with mortality, however. There was also an accident and emergency department effect.

Bentham used the simplest measure of accessibility to accident and emergency services: whether or not the local authority area contained an accident and emergency unit. On average, local authority areas with no major accident services had accident death rates higher by at least four deaths per 100,000 population than those with services (the national average rate was 39.4 deaths per 100,000 males aged 15–24 per year). This difference was statistically significant and was in addition to and independent of the effects of car ownership and social class. Bentham acknowleged the imperfections of this study, based as it was on comparatively crude measures and large areas, but the results are consistent with North American work on the links between accident fatalities and the accessibility of emergency services (Brown, 1979; Brodsky and Hakkert, 1983) and lend support to the hypothesis that a consequence of centralised accident and emergency services may be higher death rates in the remoter districts. Only much more detailed studies will provide the decisive conclusions that are needed.

ISOLATED RURAL AREAS

The problem of the physical inaccessibility of health services is not exclusively rural in incidence. It is felt by people with low levels of personal mobility in urban areas as well as in country districts. The severity of the problem, however, does increase as population density falls and services become more widely spaced. In the remotest rural areas, where population density is at its lowest and urban centres are very distant, people must cope with journeys to health services that would be unacceptable elsewhere. In much of northern Scotland and parts of rural Wales and northern England the nearest accident and emergency centre is more than 50 miles away, and in some places it is more than 100 miles distant. A trip to hospital by public transport for an out-patient appointment or to visit a sick relative may well take two days, with an overnight stay in between. Occasionally bad weather may make the journey impossible, especially if a sea crossing is involved. There is usually no choice of general practitioner in such places, where the distance between surgeries sometimes exceeds ten miles. Visits to the dentist may be most conveniently arranged when staying with relatives who live closer to a surgery. These encumbrances are a scale of magnitude removed from those experienced by the majority of rural people, yet they are accepted as a necessary accompaniment of life in a sparsely populated area. The possession of a household car has a considerable effect on whether or not access to services is seen as a problem, as Lumb (1983) found in her study of remote rural Northumberland.

The Western Isles

The Highlands and Islands of Scotland are the most isolated areas in Britain, so conditions there deserve mention. Health care in north-west Sutherland, a rugged district with only rudimentary communications, has been described by Bain (1983). The most extreme isolation, however, is to be found in the Western Isles, where rough seas are an additional barrier between people and services. An examination of access to health services there, by Bloor et al. (1978) was commissioned by the Scottish Consumer Council. Bloor and his colleagues divided the Western Isles into four categories which reflected increasing difficulties of access. Stornoway, the largest centre of population on the island of Lewis, is the best served with a small hospital containing limited specialist departments, a health centre and resident dentist, chemist, optician and chiropodist. The island of South Uist falls in the second category, with a resident dentist and a partnership of doctors who run a small cottage hospital. Other areas with resident single-handed doctors who operate main and branch surgeries are the third category, and the small islands which have no resident general practitioner are the fourth.

Outside Stornoway, Bloor *et al.* found that 41 per cent of their sample of Hebrideans lived more than ten miles from a main surgery. The journey to surgery involved hiring a car for many people and a ferry crossing for people on the smaller islands. Branch surgeries were held by all but the Stornoway doctors, and this had the effect of reducing the proportion of people more than ten miles from any kind of surgery to 14 per cent. Home visiting, however; did not appear to be compensating for the greater Hebridean distances. Bloor and his colleagues recorded home visits as 13 per cent of all consultations for children under 11 and 45 per cent of consultations for people aged 70 and more. While the age categories are not strictly comparable, these figures are below the equivalent values for national samples recorded in the General Household Survey. Accidents and emergencies are potentially most dangerous in the small islands. In 5 of the 17 emergencies recorded here, the doctor took more than eight hours to arrive after the decision to contact him.

The difficulties of providing general medical services in the Highlands and Islands have been described in the Birsay Report (Scottish Home and Health Department, 1967). In the judgement of this committee, the professional isolaton of the doctor is the main barrier to medical efficiency. Family doctors practising single-handed in remote areas are likely to be cut off from colleagues, yet meet only a light load of medical duties: a combination which matches less experience with more responsibility. The lack of choice can also be a serious problem for both doctors and patients. When only one doctor is available, both sides must make the best of doctor—patient relationships that may be less than ideal. Additional difficulties for general practitioners in remote locations include the

Table 6.11: Use of district nurses in the Outer Hebrides (%)

Use	Stornoway	South Uist	Places with single GPs	Islands with no GP
Nurse consulted before doctor:				
Elderly patients	10	7	43	85
Children	3	0	30	100
Only nurse consulted about some conditions:				
Elderly patients	7	10	33	35
Children	23	20	43	75
Nurse visits regularly:				
Elderly patients	13	20	54	70

Source: Bloor et al., 1978, p. 52.

problem of getting a locum to cover the practice at holiday times or in case of illness and the shortage of secondary schools for the doctor's own children.

Each of the inhabited small islands with no doctor does have its own district nurse, and these nurses act to some extent as 'bare-foot doctors'. They make preliminary diagnoses, treat emergency cases when a doctor is not available and supply antibiotics and analgesics at the doctor's request and on his or her behalf. The inhabitants of the islands with either no resident doctor or only a single doctor make much more use of the district nurse service than residents of Stornoway and South Uist, where there are group medical practices, as Table 6.11 shows. Whether for elderly patients or children, district nurses tend to be consulted before the doctor and instead of the doctor, to a greater extent in the more remote islands. Nurses also visit more in the areas least accessible to general practitioners.

The use of nurses as the point of first contact for health care is controversial in a service which under most circumstances would insist that a trained medical judgement is necessary to diagnose potentially dangerous conditions at an early stage, but the alternative in the smaller Western Isles would un-doubtedly be poorer access to health care. The district nurse takes on the task of reassurance and makes it easier for people to decide whether to consult the doctor (only 10 per cent of people in the small islands reported that they had difficulty in deciding whether to contact a doctor, compared with 33 per cent in Stornoway). Nurses are expected to report all cases in

177

which they have been the first contact to the appropriate general practitioner, so the general practitioner's judgement is used as well as their own in deciding which people need medical attention. Their work in providing an additional source of health care advice as well as in reporting symptoms to general practitioners helps to fill what would otherwise be a serious gap. There is therefore an argument that such work done by district nurses over and above their nursing tasks should be officially recognised and that the formal rules governing their duties should be amended accordingly.

Outside Stornoway other primary health care services are not readily available. Each general practitioner keeps a small stock of drugs, but the range is limited and when a drug is not available locally a prescription is sent to the mainland, which entails a delay; the patient must also pay the postage. Self medication for minor conditions is not easy as people must guess their requirements in advance, when they are on an infrequent shopping trip on the mainland. Several opticians visit from the mainland, holding sessions in local hotels and church halls, advertised in the local paper. The frequency and timing of visits depends on their work schedules and weather conditions (because of the ferry crossings). The outlying communities are also served occasionally by visiting or part-time dentists, an unstable arrangement which makes continuity of care difficult to achieve. Many people make appointments with a dentist when they are visiting relatives on the mainland (most usually in Glasgow) or order denture repairs by post.

Although the islanders appeared to accept their limited access to dental and optical services with stoicism, they were much more concerned about the availability of chiropody, which was not obtainable outside Stornoway, although theoretically part of the NHS. Bloor's survey revealed as many people without chiropody treatment who said they needed it as people who were receiving it, with the highest proportions in the more remote locations. Many respondents with foot problems did not seek treatment because the journey to Stornoway was too difficult or painful.

The most difficult journeys, of course, are to hospital. Patients needing specialised in-patient or out-patient treatment must usually travel to the mainland, some 90 miles to Inverness or 150 miles to Glasgow. Patients and necessary escorts can reclaim travelling expenses (although three-month delays in receiving reimbursements were reported), but overnight accommodation cannot be claimed. Regular visits to out-patient clinics or to see an in-patient can therefore impose real financial burdens. Several survey respondents referred to the grief of bereavement being made worse by the fact that the patient had died unvisited. There was also some evidence of reduced use of hospital services. Children in the Western Isles sample had a low rate of out-patient attendances, only a little more in two years than the rate in Britain as a whole for a three-month period, but the in-patient admission rate for children appeared to be comparable with the national figure. The elderly had low rates for both out-patient attendances and in-patient admissions.

Remoteness and need

For the health service, any large area with a low population density is likely to be treated as 'remote'. Although the accepted norms of so many hospital beds per thousand patients or so many patients per general practitioner might be relaxed as a matter of policy in such areas, no attempt is made to make the distances between homes and services comparable with those in the rest of the country. Through the application of many decisions about the viability of individual service locations under the constraints of fixed budgets, a 'trade-off' has developed between the principle of equal access and the expense of supplying services above the level of national norms. As Bloor et al. (1978) have pointed out, this trade-off is a political question. The principle of equality of access to health care is denied by many acknowledged barriers: difficulties in defining and recognising ill health, difficulties in making an appointment and in making the journey, difficulties in communicating with the doctor, and so on. Geographical access is only one barrier among many, and it is not necessarily the most difficult to remove, as the high level of medical facilities for the oil rigs demonstrates. Bloor and his colleagues see remoteness simply as 'where there are problems of access which it has not been thought worthwhile to overcome' (Bloor et al., 1978, p. 28).

For people living in sparsely populated areas, felt needs for health care at least partly reflect the remoteness of the service to them. There is no evidence that these people suffer any less objectively defined morbidity than people living in more central locations: indeed, the evidence appears to be in the opposite direction. In the Western Isles, for example, Bloor found higher proportions of the elderly reporting long-standing illness and restricted activity than occurred nationally. Yet the Hebridean population was found to have different expectations of health care to those of the majority of the British population, exemplified by the low value the island elderly put on teeth. People of the Western Isles seemed not only careless about the preservation of their natural teeth but also about their false teeth, for more than a third of those with false teeth did not wear them. Bloor *et al.* point out that, rather than improving their teeth, older people in the isles often preferred to adjust the uses to which the teeth were put. The cosmetic value of teeth as an aid to the appearance appears to carry little weight in a society where there are few strangers to impress. Such independence and scepticism about health values more usual on the mainland could not be traced to constraints on information. The researchers found no evidence that remoteness in the Hebrides diminished knowledge of what is technically possible in health care or claimable from the National Health Service. The demands made of a remote health service, however, are different to those made when the service is accessible. The remoteness of health services appears to have a direct effect on the perception of need.

POLICY OPTIONS

In the most isolated rural areas, a practical solution to some of the everyday problems of access to health care has already evolved unofficially and it would be appropriate to recognise it. The much greater responsibility undertaken by district nurses in such areas where medical cover is thin deserves official acknowledgement and proper reward. Encouragement might be given in the form of special training and a code of practice which gives statutory support for nurses undertaking selected tasks normally performed by doctors. Other aspects of primary care might be strengthened in isolated areas by encouraging the development of mobile services for chiropodists in particular, but also dentists and opticians.

People living in isolated areas have particular problems in visiting patients in hospital and much distress might be relieved by making available financial assistance towards travel and accommodation (as is already done for the families of people in prison). Being able to visit close relatives in hospital should cease being regarded as a privilege and be recognised as a right. The relatives of long-term patients who must travel long distances, in particular, should be considered as worthy of receiving some assistance. Whenever possible, the parents of children in hospital should be offered accommodation in the hospital itself.

For the less isolated rural areas, which nevertheless have lost both primary health care and hospital services through the trends towards centralisation described in Chapters 3 and 4, some restoration of local facilities might be possible without undue cost. In Chapter 4 it was argued that small local hospitals might have a distinctive and valuable role in the future. These would facilitate visiting for certain types of patient and help to maintain patients' contacts with community life. There is also a strong case for moving some out-patient clinics out of the urban-based district general hospitals and into smaller local centres. Partial decentralisation and an avoidance of bunching appointments for 9.00 a.m. would benefit many rural residents who are referred to consultants.

For accident and emergency services, the clarity of the argument for current policy has been clouded by evidence of deleterious effects. More research is needed to establish the nature of the trade-off between large and efficient accident centres on the one hand and the deaths of accident victims occurring due to long journeys on the other. The Platt Report, on which current policy is based (DHSS, 1962), did not contain any evidence on this issue. As such evidence becomes available, the objective of concentrating accident and emergency services in a few centres should be reviewed. In the meantime, more effective medical help for accident and emergency victims faced with long ambulance journeys might save lives. Paramedical training for ambulance crews would have obvious benefits. The comparatively dense network of general practitioners in rural areas might also be used to advantage. With radio contact, general practitioners in sparsely populated areas might reach the scene of an emergency and be able to give medical aid before the arrival of an ambulance.

General practitioners already operate such schemes on their own initiative in a few areas.

Another serious issue is how to loosen the constraints on access to non-emergency primary care in rural areas of moderate isolation. Both general practitioner and hospital services are at present rationed by the difficulties of physical access, but the available evidence suggests that the main effect operates on general practitioner consultations. The general practitioner is a key figure who deals wth most episodes of illness and who controls access to other primary health services, out-patient services and, ultimately (through out-patient referrals), in-patient beds. Inadequate access to general practitioners is therefore likely to be reflected in differential use of other health services. Improving physical access to general practitioners might be expected to change the geographical pattern of use for secondary as well as primary health care.

There are several ways in which general practitioner services might be encouraged to develop in order to overcome accessibility problems. One is to extend home visiting by doctors. This would require the reversal of a well established national trend. There are not even any signs that doctors practising in the more rural areas are sympathetic to the idea that increased visiting on their part would ameliorate a serious problem. In the Norfolk survey referred to earlier, there was no evidence of higher home visiting rates in the villages with no surgery (Bentham and Haynes, 1985). Probably only an intervention in the form of greater financial inducements for home visits would be effective.

A second way of making doctors more accessible to people with low personal mobility is to encourage contact by telephone. There is some indication from the Norfolk survey that the telephone already helps to fill the gap created by inaccessibility. Comparing the number of telephone consultations with all other consultations, there was a statistically significant increase from the villages with a surgery to those without a surgery. Telephone consultations may not be a satisfactory substitute for personal contact between patient and doctor, but increases in telephone availability among the elderly and those with low incomes would facilitate the early communication of health problems by those most at risk.

A third option is to counteract the inclination of doctors to close branch surgeries in outlying areas by a policy initiative to encourage the provision of simple but adequate part-time surgeries in settlements which have no main surgery. The available evidence suggests that the presence of a branch surgery in a village keeps general practitioner consultation levels up to those typical of villages with a main surgery. Branch surgeries with simple facilities are generally avoided by patients with high personal mobility, but they are used by the old, those without cars and people in lower socio-economic groups. Preserving even the poorest branch surgeries from closure is an attractive policy because it would benefit the most disadvantaged groups to the greatest extent.

What is perhaps the most fundamental problem of access to health care in rural areas (the lack of personal mobility for a vulnerable minority) is, of course, outside the control of the health service. The difficulties some people experience in getting to health services are part of a much wider problem. Rising car ownership in rural areas has undermined almost all village services, not just those concerned with health, together with the public transport network necessary for some people to reach them. In this situation, piecemeal policies restricted to a particular set of services are less likely to make a significant impact than a broader-based approach.

Moseley (1979) has discussed the four groups of options open to policy makers to make all services more accessible to immobile rural residents. Firstly, there are various ways of improving the transport links from people to facilities. The decline of scheduled bus services might be resisted by developing off-peak uses, better co-ordination with other agencies, initiatives in marketing and publicity, vehicle design and fares policies. More flexible public transport schemes such as 'dial-a-ride' might be tried. Combining passenger transport with a delivery service, as in post buses, and encouraging local voluntary community bus initiatives offer the possibility of public transport where none exists at present. Voluntary car schemes such as that advocated by Lumb (1983) would also fill a need. Secondly, mobile services might be used to make facilities available close to people's homes. A mobile chiropody service seems particularly appropriate, along with mobile post offices, shops, libraries and information and advice centres. Thirdly, both people and services might be concentrated together in 'key villages'. Here, planners must rely on permissive and negative powers to guide independent decisions about the location of employment, housing, roads, schools, shopping and other services within an overall long-term strategy. Finally, some services could be decentralised and established in dispersed locations by reducing the range, opening times or quality of facilities available. Branch surgeries are an example, but Moseley suggests that many facilities could be made available in rural areas on a rotating part-time basis, with 'Monday villages' and 'Tuesday villages' offering a range of services on a particular day and perhaps outside conventional working hours.

Moseley has drawn attention to the wide range of official agencies that affect the ability of rural residents to travel and the even larger number that affect their need to travel. The National Bus Company, local authority bus operators and several thousand small private firms are responsible for rural bus services, subject to regulation by the Traffic Commissioners and the Department of Transport. County councils are responsible for transport planning as well as for controlling the location of major developments through their planning departments. Social services and education are also county council concerns. District councils often administer public transport in their areas, and they determine the location of housing. Health authorities control hospitals and community health services, while family practitioner committees have some influence over the location of general medical services but less over dental, ophthalmic and pharmaceutical services. Public corporations such as British Rail, the Post Office, the Electricity Board and the Gas Corporation are also involved. While accessibility is not a central concern for most of these agencies, decisions taken by any of them have an effect on accessibility and, ultimately, on the services provided by the others. Co-ordination between them will be necessary if the forces that have produced the present situation are to be countered, and Moseley identifies county councils as the bodies best suited to organise inter-agency co-operation within an agreed strategy.

7

Health Care in the Inner City

Many inner-city areas in Britain are declining. They have several features in common. Unemployment is generally high, as firms close down and move out of the congested centres of cities. Housing is old and often in poor condition. People live at high densities and there are concentrations of ethnic minorities, single-person households and also of homeless and destitute people. Economic decline is accompanied by poverty and social problems. These, in turn, can lead to poor health.

There are also high concentrations of health services in the inner parts of large cities. Large numbers of general practitioners hold surgeries in the ample villas which surround city centres and most large hospitals are to be found not far from central business districts. On the face of it, a population with a high need for health care has all the facilities on its doorstep. Closer examination has revealed that the facilities are not always adequate.

CHARACTERISTICS OF INNER-CITY POPULATIONS

London may be taken as an example of a British city with severe problems of social deprivation in certain inner areas. Differences in the health of the population in inner London compared with outer London and England and Wales generally were investigated in a report for the Royal College of General Practitioners by Jarman (1981). Some of Jarman's evidence is given in Table 7.1, together with measures taken from the 1981 population census. The two sets of data do not match precisely in terms of dates or the areas covered (London

185

Table 7.1: Characteristics of the population of inner London compared with outer London and England and Wales

Characteristic	Inner London	Outer London	England and Wales
Percentage of households with one pensioner living alone, 1981*	15.8	13.6	14.2
Percentage of households with one person living alone, 1981*	31.8	22.4	21.7
Percentage of socio-economic group 11 (unskilled manual), 1981*	6.3	3.4	4.4
Percentage of resident population born outside the UK, 1981*	24.3	14.6	6.6
Percentage of households lacking or sharing use of a bath, 1981*	9.2	3.7	3.2
Percentage of households with more than 1.0 persons per room, 1981*	7.1	4.2	3.4
Children in care per 1,000 population aged 0–14 in 1977–8	24.3	9.8	9.5
Admissions to mental illness hospitals per 100,000 population in 1975	517	392	378
Patients treated for narcotic addiction per 100,000 population Sept./Oct. 1978	40.2	5.5	3.2
Respiratory TB notifications per 100,000 population 1976	39.3	26.1	15.9
Deaths from pneumonia per 100,000 population 1977	120	112	105
Deaths from bronchitis and emphysema per 100,000 population 1977	50	45	45
Deaths from suicide and self-inflicted injuries per 100,000 population 1977	19	13	12
Deaths from accidents (other than motor vehicle) per 100,000 population 1977	20	12	17
Infant mortality rate 1975–7	16.4	13.8	14.4

Sources: Jarman, 1981; items marked * from 1981 census (Office of Population Censuses and Surveys, 1983a).

boroughs in the census and family practitioner areas in Jarman) but the correspondence is close enough for illustrative purposes. All the main groups in the population which require high levels of health care or which have unusual health care needs are present in greater proportions in inner London than in the outer suburbs or nationally. Elderly people living alone make extremely high demands on the health service. People of any age living in single-person households and people from unskilled manual households also have relatively high health care needs. Foreign-born residents sometimes have special health care requirements. All these groups are more likely to be encountered in inner London than in outer London or in England and Wales as a whole.

Much of the housing stock in inner London is in poor condition. The proportion of households there lacking sole use of the basic amenity of a bath is almost three times the national average. Residential overcrowding is relatively high, workplaces and transport routes are congested and air pollution is worse than in the suburbs. As well as the detrimental effects of the environment in the inner city there are the selective effects of migration. Inner-city residents who move out to follow employment opportunities tend to be more active and healthier than those who remain behind. Furthermore, the cheap furnished rented accommodation available in some inner-city districts is a magnet for people from elsewhere whose personal circumstances have deteriorated. There are therefore several reasons for morbidity and mortality rates to be consistently high in the inner city. Table 7.1 demonstrates that this is the case for London.

The concentration of adverse social and environmental conditions in the inner city is not confined to London. Redfern (1982) examined the characteristics of wards in the other English and Welsh cities, as measured in the 1981 census. He found that wards suffering multiple adverse conditions outside London are mainly inner-city wards, and are much more frequent in metropolitan counties (the major conurbations) than in non-metropolitan counties. Redfern also found evidence of multiple adverse conditions in some wards in the outer areas of cities. These were typically council housing estates with high unemployment levels.

Contrasts in morbidity

Sarah Curtis (1984) has provided a more detailed examination of health contrasts in London. She surveyed two boroughs in the North East Thames Region, one (Tower Hamlets) typical of 'East End' inner London and the other (Redbridge) representative of the outer suburbs. A total of 207 households in Tower Hamlets and 237 households in Redbridge were sampled from the rating list and provided questionnaire responses on

187

Table 7.2: Morbidity in inner and outer cities

Morbidity dimension	Mean Nottingham health profile score			
	Inner London	Outer London	Inner Manchester	Outer Manchester
Energy	15.9	11.1	17.9	14.5
Pain	8.0	4.3	6.9	6.1
Emotional reactions	13.7	7.1	13.2	9.2
Sleep	17.6	15.1	16.5	14.0
Social isolation	8.6	5.1	6.7	5.1
Physical mobility	7.8	5.5	6.6	5.9

Source: Curtis, 1984, p. 89.

household characteristics, self-reported morbidity and the use of health services. Compared with the outer suburb of Redbridge, the inner-city sample in Tower Hamlets consisted of greater proportions of young men, smaller proportions of elderly women, larger proportions of single, widowed and divorced persons, more people who were newcomers to the area, more immigrants from overseas, more unemployed and more people from manual and personal service occupational classes. Households in Tower Hamlets were less likely to own their own accommodation, less likely to own cars and other consumer durables and more likely to lack basic amenities than those in Redbridge.

Morbidity levels in the two samples were measured by several indicators, including those from the General Household Survey and the Nottingham health profile (Hunt *et al.*, 1980, 1981), of which the latter were found to be the least susceptible to interviewer effects. Table 7.2 summarises the results of the Nottingham health profile test, which is a series of questions on difficulties people may have in daily life. The answers have been combined to give morbidity scores on six different dimensions. For each dimension, the inner-city population recorded consistently higher morbidity. Table 7.2 also contains the results from a similar study in Manchester (Leavey, 1983) which are remarkably similar in absolute as well as in relative terms. Curtis found significant associations between the morbidity indicators and socio-economic variables such as marital status, housing tenure, occupational class, car ownership and foreign origin. These associations were interwoven with the strong effects of age on morbidity, but they remained after variations in age had been controlled. The inference is that different levels of health in inner- and outer-city areas can be traced back to socio-economic and other differences between the two populations.

Differences in treatment

Studies of medical treatment may also be used to demonstrate differences in morbidity between populations, but they must be interpreted with circumspection. Jarman (1981) made a comparison between part of inner London (the 'West End') and outer London by examining the prescriptions of general practitioners and their claims for item-of-service payments. He found that, compared with their colleagues in outer London, general practitioners in the West End claimed fewer payments for vaccinations and immunisations, cervical cytology tests, night visits and maternity medical services. The NHS prescriptions dispensed in the West End were for particularly high proportions of central nervous system stimulants, appetite suppressants, vitamins and, to a lesser extent, hypnotics. Jarman did not comment on these differences, which may partly be due to the particular pattern of morbidity in the West End but also might reflect something of the unusual characteristics of that area's general practitioners. Jarman's description of general practice in inner London was used in evidence in the Acheson Report, to which we now turn.

THE ACHESON REPORT

The most influential investigation of primary care in an inner-city area was carried out in London by a study group set up by the DHSS under the chairmanship of Professor E. D. Acheson (London Health Planning Consortium Study Group, 1981). Inner London was defined for the purposes of this study as the parts then covered by four area health authorities: Camden and Islington; City and East London; Kensington, Chelsea and Westminster; and Lambeth, Southwark and Lewisham. The Acheson Report considered the whole range of primary health services of inner London in the context of other supporting services. In the late 1970s, inner London lost about 1,000 hospital beds and five accident and emergency centres. Further cuts in hospital services were imminent. At the same time, the social services financed by local authorities were contracting due to shortage of funds. Both trends have continued since then and have placed increasing pressure on the primary health services, as the Acheson Report foresaw.

Acheson found that the general practitioner and community health services in inner London were in no condition to bear the growing additonal burden. His report focused most attention on the general practitioner as the key figure: the first point of contact for most people and the link between other services. General practitioners' services did not themselves appear to be in good health in inner London.

Characteristics of general practice

The attributes of general practice in inner London are compared with those elsewhere in Table 7.3. It is well established that general practitioners working together in groups are better able to supply a continuous and accessible service than single-handed practitioners, yet the majority of inner-London practitioners were not in group practice. In Kensington, Chelsea and Westminster as many as 79 per cent were not in group practice. The attachment of community nurses to practices is a crude indication of co-operation between general practitioners and community nurses, but attachment rates were low in inner London. The employment of nurses by practices was also less than half the national level. High proportions of inner-London family doctors were elderly (Figure 7.1). In Kensington, Chelsea and Westminster almost a quarter of practitioners were over 65. Some general practitioners were very elderly; there were 21 over the age of 80 practising in inner London. General practitioners are not required to retire at any age. A strategy of maximum financial advantage is for practitioners to nominally 'retire' after 60, which qualifies them for continuing superannuation payments, and then resume paid NHS work with a small list of patients, covered in off-duty time by deputising services which are readily available in inner London.

Although the average list size in inner London (2,151) was not far short of the national average at the time (2,277), these

Table 7.3: Characteristics of general practitioners in inner London and elsewhere (%)

Characteristic	Inner London	Outer London	England and Wales
GPs not in group practice	59	48	28
Practices with attached community nurses	25	59	68
Practices with employed nurses	11	19	24
GPs over 65 years old	18	11	6
GPs not born in Great Britain	47	42	26
GPs with lists less than 1,500	17	10	7

Source: London Health Planning Consortium, 1981, pp. 20–5.

Figure 7.1: General practitioners aged 70 years or more in inner London, 1977

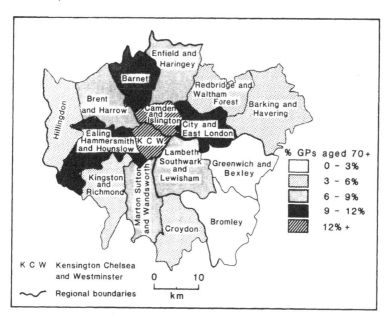

Source: Jarman, 1981, p. 61.

figures mask the high proportion of doctors with very small lists in inner London, balanced by a few practices with very large lists. While 17 per cent of doctors in the inner area had lists of less than 1,500 patients (compared with the national figure of 7 per cent), the proportion reached 28 per cent of

191

doctors in Kensington, Chelsea and Westminster. Some of the doctors with small lists are elderly and semi-retired, but others keep a small NHS list to subsidise their private work, for which there is much scope in London. Doctors with 1,000 patients on their NHS list receive full payment of practice allowances, while those with less NHS patients receive benefits on a sliding scale. Even those with as few as 100 NHS patients may be given full reimbursement of their rent and rates and 70 per cent of the cost of employing a receptionist. It is therefore much more profitable for doctors practising privately to have a small list of NHS patients as well.

Small average list sizes in inner London make it difficult for new practitioners to move into the area because average list size is the basic criterion used by the Medical Practices Committee in deciding whether a particular area has too many or too few doctors. At the time of the Acheson Report, out of the 34 practice areas in inner London, 14 were classified as restricted and 18 as intermediate. Only 2 were classified as open. Since new practices are rarely allowed in restricted areas and only in exceptional circumstances in intermediate areas, almost the only way for a doctor to start practising in inner London is to wait for a vacancy to occur. Given the reluctance of doctors to retire, this often means waiting for a general practitioner to die. Even then, young doctors are at a disadvantage, for the Medical Practices Committee gives preference to candidates with the most experience.

Ease of contact

To the inner-city resident, the age or list size of a general practitioner is of less interest than the quality of service offered. Quality of care is not easy to measure directly, but one indirect index is the ease with which the doctor may be contacted. As part of the Acheson investigation, the London Health Planning Consortium commissioned a survey which telephoned a sample of 300 practices in inner London and 150 practices in outer London at their advertised telephone numbers (as if a patient were trying to contact a doctor). The telephone was answered by a receptionist or a doctor in only 57 per cent of inner-city practices. The figure for outer London was 77 per cent. In the remaining 43 per cent of practices, the telephone was not answered at all (particularly in single-handed practices), or there was a recorded message or a GPO intercept message with instructions to call another number. This was during the working day. During evenings and weekends a higher proportion of practices could not be contacted.

Doctors' answering machines and GPO intercept services frequently directed patients to commercial deputising services. These services are available to take over a doctor's patients temporarily to allow him or her to take time off. Consent to use deputising services is granted individually to general practitioners by family practitioner committees, usually on a strictly limited basis. A single-handed doctor, for example, might expect to be allowed to use a deputising service for two nights and one half-day per week. Although they are staffed by qualified doctors, commercial deputising services lack continuity and cannot be expected to provide the same quality of care as an experienced family practitioner. The Court Report, for example, was critical of the increasing use of deputising services in Britain from the point of view of child health. It was seen as a retrograde step for parents who call their doctor after a period of anxiety about their child's health (and often after a night of distress) to find themselves consulting a doctor who is a stranger to the child and to them (Committee on Child Health Services, 1976).

According to the Acheson Report, 39 per cent of general practitioners in England and Wales had been given consent to employ a deputising service. As many as 98 per cent of the general practitioners in the City and East London area and 73 per cent in Kensington, Chelsea and Westminster were approved users. The widespread use of answering services, answering machines, intercepted calls and redirected calls made it difficult to tell whether at any time doctors were being covered by a deputising service or not. The Acheson Report was not confident that family practitioner committees were always aware of the extent of use of deputising services in their areas, and mentioned the possibility of abuse. The evidence available suggested that in some practices all out-of-hours cover is provided by deputising services.

Surgery accommodation

Another crude indicator of the quality of care patients receive is the quality of the surgery premises in which they are examined and treated. The Acheson Report tells a sorry tale on this subject also. Only about one-quarter of general practitioners' premises in inner London came within the advisory standards recommended by the DHSS. These standards include provision for other health care staff, which was rarely made. More disturbing was the estimate made by regional medical officers that about 15 per cent of surgeries (and 30 per cent in some areas) were not fit to be used for the provision of general medical services. Some of the worst surgeries occupied lock-up shop front premises with little or no heating or sanitary facilities. While it is the family practitioners committees' responsibility to see that 'proper and sufficient' accommodation is provided, there were no minimum standards and no systematic inspections of premises by family practitioner committees in London.

The problem of many unimproved surgery premises in inner London arises from a combination of poor-quality structure and high site values. Suitable premises are hard to find and the conversion and improvement of unsuitable premises are not likely to be a worthwhile property investment. Doctors' surgeries are worth less on the open market than dwelling houses or shops. As far as running costs are concerned, the indirect reimbursement of practice expenses (heating, lighting, furniture, equipment and so on) benefits those who provide sub-standard accommodation to the detriment of those whose better-appointed premises are more expensive to maintain.

Linked to the problem of sub-standard surgery premises in the decaying environment of the inner city is the preference of many doctors to live and raise their own families in more salubrious and perhaps less expensive surroundings at some distance away. In the City and East London area, for example, less than one-third of doctors lived in the same borough as their surgery location.

194

Table 7.4: Characteristics of health visitors in inner London and elsewhere (%)

Characteristic	Inner London	Outer London	Home counties
In post less than 2 years	53	37	35
Less than 30 years of age	32	19	13
Unmarried	55	25	21
Working part-time	14	27	24
Recruits qualified in previous year	73	68	50
Staff leaving during year	27	21	14

Source: London Health Planning Consortium, 1981, pp. 61–2.

Community nurses

When the Acheson study group turned its attention to other health services in inner London, it found that these were often not strong enough to make up for the deficiencies in the general practitioner service. There are serious problems in inner London in recruiting and retaining community nurses. Jarman (1981) noted that in some inner-city areas the community services budgets were underspent because of the difficulty of attracting nurses, health visitors and other staff. Table 7.4 summarises the characteristics of health visitors in inner and outer London north of the Thames (the survey refers to NE and NW Thames Regional Health Authorities). Compared with their counterparts in outer London and the Home Counties, health visitors in inner London were younger, less experienced and had fewer ties with the area (they were less likely to be married or working part-time: part-time working was not encouraged). They were also more likely to leave. The figures for district nurses showed the same profile. Many stay only for a short time because of both living and working conditions. It is difficult to find suitable living accommodation. Community nurses in the inner city have high workloads not only because the population there is socially deprived and has high health care needs but also because nurses find themselves trying to make up for deficiencies in other services. District nurses, for example, may be doing jobs which ideally the personal social services should take care of. Community midwives are in a particularly difficult situation when they are responsible for pregnant women who are unable or unwilling to register with a general practitioner. Statutory care must be given in these circumstances without medical back-up.

Problems of co-ordination between general practitioners and community nurses were also highlighted. The considerable overlap in practice areas meant that attachment of community nurses to general practitioners was not an ideal response because it was often wasteful in terms of visiting time. It could also lead to other difficulties in coverage. Community nurses are employed by district health authorities and are not allowed to cross district boundaries, whereas general practitioners' practice areas frequently do. Furthermore, the people not registered with a general practitioner (who could amount to almost one-third of the population in some areas) would be in danger of being missed if community nurses concentrated on the patients on particular lists. But whether community nursing is organised on an attachment or a patch basis, the business of regular meetings between the different providers of primary care in an area (including social workers) remains physically difficult. Few general practitioners' premises have accommodation for other primary care staff, so there is often no suitable place in which the 'team' can meet.

Non-registration

The area designation scheme used by the Medical Practices Committee to make the geographical distribution of general practitioners more even is based on the assumption that doctors with small lists will be underworked and will be trying to get more patients to increase their incomes. The evidence considered in the Acheson Report suggests that this assumption is not true in inner London. Many inner-city general practitioners had effectively closed their lists and were not interested in taking more patients. Two surveys were conducted by the Soho and Marylebone Community Health Council, in which all practices in the council's area were telephoned and asked to accept a young married couple with three children on to their list. On both occasions, two-thirds of the practices did not co-operate. After considering this and other evidence, the Acheson Report came to the conclusion that one reason for non-registration with a general practitioner is the difficulty of finding a general practitioner willing to take new patients.

There are other reasons, of course. In a mobile and varied population there is always a proportion of people who have not yet tried to register, who do not know how to or who do not wish to register. Although only 2 per cent of the national population is not registered with a general practitioner, the Acheson Report quotes an estimate that the proportion may be 30 per cent in one inner London district. The problem is by no means uniformly serious, as registration levels in Tower Hamlets, for example, are high and similar to those of an outer London district (Curtis, 1984).

Allsop and Lovell (1984) have investigated non-registration in the London Borough of Brent. They collected information by sending a postal questionnaire to people who had attended an accident and emergency unit and said they had no general practitioner. The replies they received are not necessarily representative of the group of people who are not registered with a doctor as a whole. This is because only those who had sought medical help (in hospital) were included and, of these, only one-third replied. Nevertheless the replies are of interest as it is difficult to find out anything about such an anonymous group.

The unregistered people who contacted Allsop and Lovell tended to be young (16−30), male and highly mobile, often with several recent changes of address. The main reasons they gave for not registering were being refused after trying to register, not trying to register because they were 'never ill' or because they preferred to use an accident and emergency unit, and not trying to register through not knowing how. Hardly any had seen a list of doctors practising in their area. The researchers checked the local family practitioner committee's claim that the list was available in post offices, chemists, police stations and libraries and found that only local libraries could be relied upon to have a copy. Family practitioner committees, it seems, are sometimes not active in pursuing the interests of patients or potential patients because their main role is in working with doctors and other health care professionals.

Use of accident and emergency departments

What happens to the people who are not registered with a general practitioner when they become ill? Some of them find their way to deputising services, even though these are not advertised. Others present themselves at accident and emergency departments in hospital. The Acheson committee commissioned a survey of patients attending accident and emergency departments in inner and outer London and found that 25 per cent of patients in inner London said they were not registered with a local general practitioner, compared with 9 per cent in outer London. A later survey in London by Farmer (1984) found as many as 40 per cent of accident and emergency unit patients said they were not registered with a local general practitioner.

Although a substantial minority of accident and emergency department users in the inner city are not registered with a general practitioner, the majority of them are, but prefer to use the hospital for primary care. In the Acheson survey more inner-city residents than outer-city residents said they had used the hospital because they felt their own doctor would not be available or provide the necessary treatment. Some had seen their own doctor first but still found it necessary to go to the accident and emergency centre. This proportion was higher in inner London (72 per cent) than outer London (50 per cent). Accident and emergency departments are used to a considerable extent as substitutes for general practitioners in central London, both for registered and unregistered patients.

Children make especially heavy use of hospital casualty departments as an alternative to consulting a general practitioner. Figures from the Children's Hospital in Sheffield suggest that some 20 per cent of all children living in inner-city areas attend in the course of a year. One-third of this enormous number are under 3 years old (Committee on Child Health Services, 1976). Most general hospitals do not aim to provide a primary care service. In accident and emergency units patients are treated by casualty officers used to dealing with accidents but often very inexperienced in general practice and especially in child health problems. There is a danger that treatment given in hospital may be inappropriate.

CONDITIONS OUTSIDE LONDON

Most investigations of inner-city health and health care have concentrated on London. It would certainly be unwise to extrapolate the findings of the Acheson Report to all large British cities. Conditions in London might well be exceptional. Inner Manchester, for example, appears to have few of the problems described. The Manchester Family Practitioner Committee has a policy of routine visits to inspect surgery premises, as the Acheson Report itself reveals. Wood (1983) looked for differences in the general practitioner service between inner Manchester–Salford and the outer suburbs of Greater Manchester and found few. The inner area did have higher proportions of elderly doctors, doctors trained overseas, single-handed practices and doctors with relatively small lists, but the differences between inner and outer areas were not as great as in London. The organisation of general practice was broadly similar in the two areas. The range of practice staff employed, the clinical equipment and record-keeping procedures, the use of appointment systems and access to support services in hospitals showed no pronounced differences. The use of deputising services was also similar, as was the attachment of health visitors and district nurses, although the levels reported throughout the city may cause some concern. In the inner area 96 per cent of doctors used deputising services (especially for night calls) compared with 98 per cent in the outer area. A sizeable proportion of doctors (19 per cent in the inner and 16 per cent in the outer area) said they had no regular contact with both district nurses and health visitors.

The use of services by patients in inner and outer Manchester is also similar, according to Leavey (1983). In spite of the inner Manchester population having higher rates of unemployment, poverty and mortality than the suburban population, Leavey could detect only a slight increase in self-reported morbidity in the inner area. Inner and outer Manchester residents were almost equally likely to be registered with a general practitioner, to have consulted in the last six months, to have received a home visit or to have attended a hospital accident and emergency department. If there are particular problems in health care associated with the inner city in Manchester, investigations so far have failed to find them.

There are, nevertheless, some indications that inner London is not unique in its health care problems. The intensity of unemployment, housing stress, environmental deprivation, social problems and ill health in parts of Glasgow, for example, certainly rivals that of inner London, and Glasgow's general practitioners are similarly elderly and single-handed in the worst areas (Knox, 1979b).

Description of general practice conditions in inner-city areas in Liverpool, Glasgow, Manchester and Birmingham may be found in Bolden (1981). In this personal account by a general practitioner, the portraits of dilapidated practice premises in Liverpool and Glasgow are memorable. Three doctors in inner Liverpool, for example, were occupying totally boarded-up shop premises and the boarded-up terrace house next door. There was a hole in the wall between the two premises which enabled the doctors to call themselves a group practice and claim the appropriate salary allowance. Liverpool's 'Harley Street' was a row of decayed terrace houses (Boundary Street) standing alone in a wasteland of demolished sites. Only one house was occupied, two were public houses and five were empty and derelict. Of the remainder, four houses were used as general practitioner surgeries (one with the windows bricked up), one was a chemist (also bricked up) and the last was a dentist's surgery. In Glasgow, vandalism struck Bolden as one of the main problems for doctors. One community clinic was broken into every day of the week he visited it. Some surgeries on council estates were described as small concrete buildings protected by barbed wire. Nearer the centre of Glasgow some doctors were moving their surgery premises from one condemned property to another as demolition proceeded. Bolden speaks of 'left-wing' local authorities being unsympathetic to the needs of doctors in not helping by providing suitable land or premises and, in some cases, planning permission. He acknowledges, however, that some inner-city doctors have never improved their sub-standard premises (often of the lock-up type) despite financial opportunities to do so.

Bolden found widespread use of deputising services for night calls in the inner-city areas he visited. A minority of doctors, he reported, made maximum use of deputising services. These general practitioners were never available except in very restricted office hours. Some doctors did not employ a receptionist or secretary but many more did not appear to be using the services of district nurses and health visitors to full advantage. A health visitor in a deprived part of Liverpool, for example, was finding it difficult to liaise with up to 30 general practitioners in her patch. Communications between general practitioners and consultants in the large inner-city hospitals were also far from ideal.

POLICY ISSUES

The Royal Commission on the National Health Service (1979) gave particular attention to the problems of declining urban areas, concluding that the NHS is 'failing dismally' to provide an adequate primary care service in some areas and in parts of London in particular. The Commission identified improving the quality of care in declining urban areas as the most important problem facing NHS services in the community and made a number of recommendations, most of which were taken up in the Acheson Report.

Improvements in general practice

Acheson's most controversial recommendation was that general practitioners should retire at the age of 70. This would need to be a national policy rather than a local one and compensation would have to be paid, but compulsory retirement would create the large number of vacancies required for the reorganisation of practices. All other health service workers retire at or before 65, but general practitioners have since made it clear that they would vigorously defend their present privilege if Ministers were to show signs of following Acheson's suggestion.

To encourage young general practitioners to move into inner-city areas, the study group recommended that they should be offered two-year posts as assistants to elderly practitioners, with the prospect of taking over the practice when the principal retires. They should then be actively helped to team up with other young doctors, for instance by the health authority finding suitable premises. To compensate doctors for the higher workloads associated with some inner-city areas it was suggested that general practitioners under 65 working in group practice in 'underprivileged areas' (yet to be defined) should receive group practice payments 50 per cent higher than the national rate.

A combination of incentives and penalties might be used to increase general practitioners' list sizes in inner cities. Doctors could be encouraged to take more patients if there was a registration fee payable when a new patient registers in addition to the existing capitation fee. A registration fee would compensate doctors for the short-term increase in workload expected with new patients. Doctors might be slightly discouraged from keeping small lists to pay their expenses if full expenses were payable only for lists larger than 1,500 patients, rather than the threshold of 1,000 at present.

Changes to the rules of classification followed by the Medical Practices Committee were suggested in order to make it less likely that areas in which people have real difficulty being accepted on a practitioner's list are still classified as restricted. For classification purposes, single-handed practitioners over 65 with less than 1,000 patients might be excluded from the calculations. Another proposal was that practitioners with lists between 700 and 1,250 should only be counted as half a doctor when calculating average list sizes.

Proposals designed to improve the quality of care provided by inner-city practitioners concentrated on improving premises, controlling the use of deputising and answering services, and facilitating more co-ordination with other members of the primary health team. The Acheson Report recommended that two-thirds of the cost of making improvements to premises should be reimbursed; twice as much as at present. All surgery premises should be inspected by family practitioner committees and a minimum standard should be set (they should have a consulting room, a waiting room, a wash basin and a flush lavatory) in order to qualify for rent and rates reimbursement. Deputising service arrangements should be reviewed regularly by family practitioner committees and doctors who choose to use them for substantial amounts of out-of-hours cover should forfeit their out-of-hours allowances. Whenever practicable doctors should be contactable at a single telephone number answered by a person rather than a machine.

Research on the extent and effect of the overlap of practices in densely populated areas was recommended. In the meantime, family practitioner committees were urged to encourage doctors to concentrate their practice areas in order to reduce overlap. This would decrease travelling time for routine and emergency visits, increase identification with a particular community and help co-ordination with other primary health care services. Where a health or local authority boundary hinders the development of teamwork between doctors and community health workers, the health workers should be allowed to cross it.

Community services

Various recommendations were made in the Acheson Report to improve the conditions of work of community nurses in inner-city areas. These included higher establishment levels, higher London weighting in salaries, better premises from which to work, more residential accommodation and two-way radios in high-risk areas. Proposals were also made to strengthen the services which make up for gaps in general practitioner care. These include the child and school health services and preventive care for the elderly. Accident and emergency departments in hospitals which undertake primary care should improve communications with primary health care teams and, at the very least, notify a patient's doctor that treatment has been given. Where possible, general practitioners should be employed in accident and emergency departments to deal with the health problems in which casualty officers may not be experienced. Community hospitals might also be developed in urban areas to serve their local communities, particularly the elderly and the mentally ill.

The Acheson Report contained 115 recommendations designed to remedy significant deficiencies in primary health care in deprived inner cities. Almost all of them required financial resources or increased regulation of the medical profession. While Ministers have not been forward in adopting them as policy, some initiatives have been made.

The DHSS response

Incentive payments were introduced in 1983 and 1984 by the DHSS to encourage group practice and the improvement of surgery premises in inner-city areas. These are both temporary schemes, operating for a limited period of a few years, and applicable in the areas which have the status of 'partnership' or 'programme' authorities under the government's inner cities policy, or which are in the area of the London Docklands Development Corporation. In addition to inner London, there are 23 urban areas in England in which the general practitioners would qualify. The first of these schemes is to assist general practitioners who wish to improve their surgeries. Normally grants of one-third are payable for surgery improvements, but this provision is for 60 per cent improvement grants in the inner cities. Single-handed practitioners are not eligible if their list size is less than 1,500 (Department of Health and Social Security, 1983e). The second scheme applies to single-handed doctors who agree to join together in group practices of not less than three doctors. Each receives an incentive payment of £4,000 in addition to the normal group practice allowance (DHSS, 1984f).

Another initiative from the DHSS has been to require each family practitioner committee to establish a deputising services subcommittee with the job of ensuring that deputising services are of a satisfactory standard and are not over-used. The committees were reminded that out-of-hours services should be of no less standard than those provided within working hours and it was suggested that they may wish to control the uses of deputising services within their areas by setting an upper limit on the number of visits per month or the amount of time covered. Whatever the local circumstances, however, family practitioner committees were instructed not to consent to any standing arrangements under which a deputising service would be employed every night and weekend (DHSS, 1984b).

Identification of underprivileged areas

Meanwhile, the British Medical Association has been marshalling the arguments in favour of greater financial incentives for general practitioners working in socially deprived areas.

Table 7.5: Census variables and the extent to which they increase the workload of general practitioners

Census variable (%)	Average score
Elderly living alone	6.62
Aged under 5	4.64
Unskilled (social class V)	3.74
Unemployed	3.34
One-parent families	3.01
Overcrowded	2.88
Changed address within one year	2.68
Born in New Commonwealth and Pakistan	2.50

Source: Jarman, 1984, p. 1590.

Jarman (1983, 1984) has demonstrated a method of identifying the areas likely to give family doctors the heaviest workloads. A list of census variables was sent to a 10 per cent national sample of general practitioners, who were asked to give a score to each variable according to their opinion of its effect on workload. The range of scores was from 0 ('no problem') to 9 ('very problematical'). The average scores recorded by the general practitioners for eight of the census variables used are given in Table 7.5. Of the census variables suggested, it was the percentage of elderly people living alone that was collectively thought to increase family doctors' workloads the most.

Jarman's method is to combine the eight census variables into a single index. Firstly, the values of each variable in the 9,265 wards of England and Wales are extracted from the 1981 census. These data are transformed to make them normally distributed and then standardised so that every variable has a mean of zero and a standard deviation of one (a common statistical procedure to make different measures directly comparable). The standardised values are then multiplied by the scores in Table 7.5 so that each variable is weighted by its relative contribution to workload. Adding the products gives the final composite score, which Jarman calls the underprivileged area score.

To check whether Jarman's underprivileged area scores tallied with the subjective impressions of local doctors, the underprivileged areas subcommittee of the British Medical Association asked local medical committees in five areas to classify wards into 'worst', 'intermediate' and 'best' categories with regard to workload. There was a high degree of

Figure 7.2: Increased workload or pressure on general practitioner services in Bradford: estimated by local medical committee and calculated by underprivileged area scores

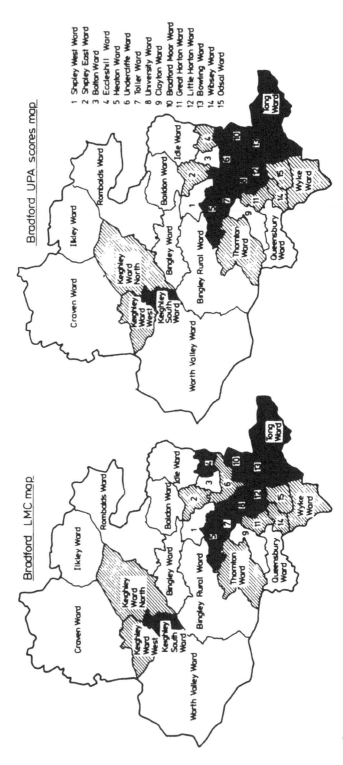

Source: Jarman, 1984, p. 1588.

Table 7.6: Jarman's top 50 underprivileged area wards

	Electoral wards (1981 census)
London 12	Tower Hamlets 3, Brent 4, Lambeth 2, Wandsworth 2, Hackney 1
Birmingham 10	Ladywood, Aston, Deritend, Duddeston, Sparkbrook, Soho, Handsworth, Newtown, Rotton Park, All Saints'
Manchester 3	Moss Side, Hulme, Miles Platting
Blackburn 2	Brookhouse, Cathedral
Bradford 2	University, Little Horton
Liverpool 2	Abercromby, Granby
Bolton 2	Derby, Central
Rochdale 2	Smallbridge and Wardleworth, Central and Falinge
Coventry 2	Foleshill, St Michael's
Nottingham 2	Lenton, Radford
Leicester 1	Wycliffe
Sandwell 1	Soho and Victoria
Burnley 1	Calder
Preston 1	Park
Wolverhampton 1	St Peter's
Calderside 1	St John's
Pendle 1	Whitefield
Bristol 1	St Paul
North Tyneside 1	No. 6
South Tyneside 1	No. 10
Peterborough 1	Central

Source: Jarman, 1984, p. 1592.

correspondence between maps produced in this way by local medical committees and maps based on a threefold division of high, medium and low underprivileged area scores derived from the census variables. Figure 7.2 gives the example of Bradford. In this map the areas shaded black are the worst and the white areas the best. These results were considered by the BMA's underprivileged area subcommittee to indicate that the method is a satisfactory way of identifying areas where, according to general practitioners' opinions, the population increases their workload or pressure on their services (Jarman, 1984).

The worst 50 wards in England and Wales according to 1981 census figures are listed in Table 7.6. This list makes it clear that, for general practitioners, populations implying high workloads are particularly associated with inner-city wards. The ward with the highest underprivileged area score is Brookhouse ward in Blackburn, Lancashire. Jarman (1983) suggests that the wards with the highest underprivileged area scores might qualify for special assistance, including the possibility of higher remuneration to compensate doctors practising in them.

'Underprivileged' has a special and limited meaning in this context. The census variables used in the measure include some that are interrelated and commonly associated with inner-city social deprivation. The unskilled, unemployed, one-parent families and overcrowding measures have correlation coefficients between themselves ranging from 0.46 to 0.67 (Jarman, 1985). (This raises the technical question of whether they should be combined in an additive fashion.) Other variables, including the elderly and children measures which are given the highest weighting, are not necessarily associated with social deprivation. Wealthy retirement areas, for example, have high proportions of elderly people living alone. Jarman's measure is an estimate of the relative demands per capita of population likely to be made on general practitioners, not a general measure of social underprivilege or deprivation. It does not indicate how doctors respond to expected increases in demand.

The connection between the needs or demands of a population and the actual workloads of general practitioners has yet to be made. Doctors practising in 'underprivileged' areas might work harder than their colleagues elsewhere, but this has not been demonstrated. The evidence from London available to the Inner London Health Consortium study group suggested the reverse. Special assistance in the form of extra money for general practitioners might increase the quality of the service in some areas but this, too, remains an untested proposition.

More radical policies

Most of the policy options for improving primary health care in inner cities involve strengthening conventional services and providing more of the same. But alternatives are possible. Several innovations in health care are being developed to meet the needs of special 'at risk' groups in inner cities (Forsythe, 1983). In Manchester there is a peripatetic general practitioner service for the homeless. A general practitioner in London has a walk-in clinic for people who are not registered with a doctor. Medical support for young unemployed people is provided in a community workshop in Manchester. London has a crisis intervention centre for homeless drug users where immediate medical, nursing and social work care is available for short-term residents. In Lewisham there is a walk-in mental health advice centre staffed by a multidisciplinary team. The centre also deals with mental illness emergencies in the community, having a psychiatrist, a nurse and a social worker available to visit homes, hostels or prisons.

Alison While (1982) believes a still more radical change of approach is needed. While found that the reason a high proportion of infants in inner London do not complete their course of immunisation appears to be connected with ethnic origin and the stability of the parents' marriage or cohabitation. She suggests that the present health care system, based on white middle-class values, may be alien to some inner-city people it tries to serve. What may help, she feels, is a system of 'barefoot doctors': health workers chosen for training by the community. These workers would be acceptable to the local people and would understand their problems, but should know enough about health to recognise problems beyond their scope.

McCarthy (1983) has also argued that we should learn from the experience of Third World countries and attack health problems in inner-city areas in a more radical way. He advocates a community development programme, similar to those introduced in some deprived urban areas in the 1960s, but with the aim of helping local groups to take responsibility for their own health. While an educational programme imposed from above has little chance of success, McCarthy gives examples from London and Paisley of how community workers participating in the life of an area can act as catalysts. The objective would be to encourage the growth of social networks for mutual support with respect to health and to promote mutual action (sometimes in the political arena) to change conditions contributing to ill health. Communities may also take responsibility for shaping local health services to meet local needs more effectively.

But health services will make little impact on inner-city health problems by themselves, however imaginatively they are reshaped. Consider, for example, a Bangladeshi resident of Tower Hamlets living with his family of seven children in accommodation with five rooms:

> This house is full of rats, damp, broken toilet, no heating, no bathroom. Children always have coughs and colds, are always scratching their bodies because they can never take a bath − no hot water system. Wife is not well because of housing conditions − uses cold water to wash clothes; can boil a kettle but [it is] not enough. Draught comes in windows because they are broken. [The] children are scared of rats and they cry. (Curtis, 1984, pp. 151−2)

Illustrations like this make the connections between ill health and poverty seem direct and obvious, but the situation is more complex. The Black Report did identify several factors such as overcrowding, nutrition, smoking and accidents at work, all of which are associated with poverty and have clear connections with disease. But these factors by themselves were not thought to be sufficient to explain the prevalence of ill health amongst the poor. More diffuse consequences of the class structure were thought to be at work as well (DHSS, 1980c). Of course, not all inner-city residents suffer from poverty or ill health. Ill health is not always associated with social deprivation and bad housing, and, where an association does exist, the question of causality is not at all clear. People who suffer from poor health tend to gravitate downwards in the economic system and so they may come to occupy the worst housing and the least attractive jobs (and take up an 'unhealthy' way of life). In such cases the correlation between illness and environmental conditions is misleading. Inner-city areas, which offer cheap and accessible accommodation, are the last refuge for many people in hardship. Policies to facilitate the delivery of health care in deprived inner-city areas do promise to give a large number of people a more equitable share in National Health Service treatment, but they cannot be expected to raise levels of health significantly. They are directed at the symptoms, not the cause.

8

Health Services for All

Previous chapters have described the geographical organisation and availability of state-provided health services. Such a viewpoint is limiting in that it takes no account of other services which overlap with those of the NHS or with alternative ways of addressing the problem of ill health. The focus is now shifted to this wider context. Some of the geographical implications of private health service development will be discussed, followed by a consideration of how certain social welfare services might meet some health needs more effectively than conventional health services. At present, social welfare services are not compensating for the uneven geographical distribution of the National Health Service in any systematic way and the private health sector is compounding the inequalities. For the NHS, a central aim remains to provide equal opportunity of access to health care for people at equal risk. Priorities for the development of NHS services and progress towards them will be reviewed in an attempt to identify the tasks ahead. It will be argued that the rationing of health services by geography remains a significant impediment.

GEOGRAPHICAL IMPACT OF PRIVATE HEALTH SERVICES

While the majority of people in the United Kingdom receive health care under the National Health Service, an increasing number elect to pay privately for consultation and treatment, either to shorten waiting times, to obtain the comforts of special accommodation, or to secure the attention of a particular specialist. These options were always open to the wealthy but they have become more widely available recently due to the growth in private health insurance. The number of people insured for health rose from about 1 million in 1960 to twice that number in 1970 (National Association of Health Authorities, 1983) and had reached 4.1 million by 1984 (Mohan and Woods, 1985).

Private health care may be obtained both within and outside the National Health Service. Within the NHS, general medical and dental practitioners are free to do as much private work as they wish, providing it does not interfere with their NHS work: that is, providing they fulfil their contracts with the family practitioner committee. Most general medical practitioners do little private work, but there are exceptions, many of them based in central London. It is more common for dentists under contract with the NHS to undertake substantial amounts of private work. Hospital consultants employed by the NHS may also practise privately, up to an earnings limit of one-tenth of their full-time salary. They may see private patients at an NHS out-patient clinic, in which case a charge is payable to the NHS. They may also admit private patients to 'pay beds' in NHS hospitals, and the patient then pays a fixed charge to the health authority for accommodation and services in addition to the fee for private treatment. There is a limit to the number of private patients in an NHS hospital at any one time, a figure in the order of 1 per cent of the total.

Other health services in the United Kingdom operate outside the NHS. A limited number of general medical practitioners (mainly located in the Harley Street area of central London) and a larger and more widespread number of dental practitioners offer only private services. Some private hospitals, such as ones managed by religious and charitable institutions, have survived as independent units from the foundation of the National Health Service. There has also been some growth in hospices for the terminally ill run by non-profit-making institutions. A quite separate development has been the increase in private acute hospitals.

The growth of private medical care

Private acute hospitals have been a recent growth industry in Britain. Their development, concurrent with the huge expansion

212

in medical insurance schemes, can be traced to the attempts of the 1974-9 Labour government to separate private practice from the NHS. The 1976 Health Services Act introduced a Health Services Board with the task of phasing out pay beds from NHS hospitals and controlling the development of private hospitals. The authorisation of the Board was required for any hospital project involving at least 100 beds in Greater London and 75 beds elsewhere. Authorisation could be withheld if the proposed hospital was likely to affect the NHS adversely. The result was perhaps the opposite of that intended: a rash of private hospital developments with fewer beds than the number required for statutory control. The growth was in the field of elective surgery for a narrow range of repair operations, such as those for hernias, varicose veins, haemorrhoids and orthopaedic, ENT and gynaecological procedures. In the five-year period until the Act was reversed, the number of private surgical beds outside the NHS grew by about 38 per cent (National Association of Health Authorities, 1983).

Private hospitals were therefore already well established when the 1979 Conservative government took office, pledged to restrain public expenditure on the welfare state and encourage private enterprise. Independent medical services were then nurtured by facilitating NHS links with private hospitals (DHSS 1981b), revising consultants' contracts to make limited private practice easier and by allowing tax concessions on companies' health insurance schemes for the benefit of employees. Under the Conservatives, the Health Services Board was abolished and its power of authorisation of private hospital development was transferred to the Secretary of State for Social Services. The minimum size threshold for referral to the Secretary was increased to 120 beds, and the Secretary was empowered to withhold authorisation only if the proposed hospital would interfere to a 'significant extent' (undefined) with NHS services. These new regulations were therefore a relaxation of the controls imposed by the previous administration. While the Secretary of State was given the power to designate districts in which all private hospital proposals would require authorisation, irrespective of the number of beds, by 1984 no such designations had been made. In the 1980 Local Government and Planning Act, the control over development of local planning authorities was reduced. Planning authorities were advised not to be swayed in their decisions by opinions on the effects of proposed private hospital developments on the NHS, and health authorities were discouraged from expressing views on the need or otherwise for private hospitals in their territories (Department of the Environment, 1981).

These reductions in control were met by a further increase in activity in the private hospital sector. Many of the charitable institutions which had dominated the sector until the rapid expansion of the market succumbed to competition from commercial companies of British and American origin with more lavish finance available. By 1984, however, the boom in private medical insurance was over. Almost all the catchments able to support a private hospital at present demand have already been identified and exploited. There may even be an excess of supply over demand. Little more private hospital building is envisaged for the immediate future, so it is an appropriate time to take stock of the geographical distribution of existing developments, as Mohan (1984b) has done. His findings are summarised below.

The location of private hospitals

Figure 8.1 gives the geographical distribution of private hospitals in Britain. A thin scatter of charitable institutions persists over the country, but most profit-making institutions are found in south-east England and the Midlands. By far the largest concentration, impossible to include at the scale of Figure 8.1, is in the West End of London. Three factors account for this pattern: the location of catchment populations, the proximity of large NHS hospitals and the availability of suitable sites.

Since private hospitals need occupancy rates of at least 70 per cent in order to be profitable, small hospitals with less than 25 beds, in which it is difficult to maintain high occupancy rates, are less desirable business propositions than larger units. For a 30-bed hospital providing treatment for 7 per cent of its insured population in a year, Mohan estimates that a catchment of 21,000 insured people is required. The geographical distribution of the insured population is therefore a critical precondition. This mirrors the distribution of wealth, with the south-east being conspicuously high. Mohan quotes an estimate

Figure 8.1: Location and ownership of private hospitals in Britain

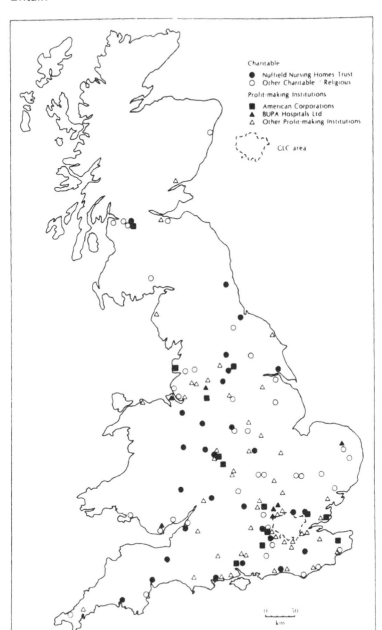

Source: Mohan, 1984b, p. 13.

of 15 per cent of the population being covered by private health insurance in the South West Thames Region, compared with 6–9 per cent in most regions outside the south-east, with the Northern Region (3 per cent), Wales, Scotland and Northern Ireland (4 per cent) at the lower end of the scale. At the local level, the presence of one or more large employers who introduce health insurance as a benefit for employees may be enough to exceed the threshold population and attract a private hospital. In Southampton, for example, the insurance schemes of Esso, Plessey and IBM attracted a new Hospital Corporation of America hospital. Another indicator of a favourable income and socio-economic setting is the investment capital available for private hospital development. Both British and American firms are said to look for substantial support for proposed hospital developments in the form of financial backing from local businesses.

Proximity to NHS hospitals is also important. Most private hospitals in Britain are not self-sufficient in terms of medical staff, but employ NHS consultants part-time. In 1980 the NHS consultants' contract was revised to allow consultants working 'whole-time' for the NHS to receive additional income from private sources up to a limit of 10 per cent of their gross salary. Part-time contracts were also revised and the attraction of additional private work has increased. Private hospitals also depend on the ancillary and back-up facilities of NHS hospitals. Of 178 private hospitals with operating theatres in England in 1983, only 102 had radiology facilities and only 54 had pathology laboratories (DHSS, 1983c), so most had some need of NHS services through the influence of consultants. Mohan describes a typical BUPA hospital as regularly using 40–50 consultants, with perhaps up to 200 consultants altogether on the hospital's medical list. There are therefore strong reasons for private hospitals not only to locate close to NHS hospitals in order to reduce the travelling time of busy consultants, but also if possible close to clusters of several large NHS hospitals, in order to maximise the number of potential consultants and facilities available. Such clusters are to be found only in the larger cities, and particularly in London.

Proposals to establish private hospitals close to NHS hospitals have usually been supported by consultants and other hospital workers because of the opportunities of additional or

216

alternative employment. Private hospitals are able to compete favourably for nurses and ancillary staff by offering attractive remuneration, more flexible working hours and pleasant modern working conditions. They also offer some non-medical jobs, which is usually seen as a benefit in declining inner-city areas with high unemployment. On the other hand, private hospitals near to large NHS hospitals in inner urban settings are of no direct benefit to the health of most local residents, and may be detrimental to the public services that are used locally. To a large teaching hospital which is already over-stretched, the use of NHS facilities for private purposes by consultants exercising their contractual right and the loss of trained staff such as theatre and intensive care workers could have a disadvantageous effect on the service for local people. Such arguments, however, are not admissible when proposals for private hospital developments are being considered. Under current planning law, the material effects of a development (its effect on jobs, traffic, the physical appearance of the buildings, and so on) can legitimately be considered by planning commit-tees, but the wider social implications and the effects of competition with the NHS are not regarded as material plan-ning considerations for the purpose of the 1971 Town and Country Planning Act.

As well as fostering essential links with existing hospitals in the public sector, private hospitals have shown some tendency to be attracted to the sites of NHS facilities no longer in use. The growth of private hospitals coincided with the closure of NHS hospitals or parts of them and also the sale of NHS land, especially the grounds of psychiatric hospitals whose retention could not be justified in economic terms, so the opportunities for redevelopment were available at the right time. The main attraction, however, was that these sites were already zoned as hospital land use, so planning permission was virtually assured.

While there has been some redevelopment of former hospi-tal buildings, there is also a strong opposite tendency to create an image of a private hospital as a modern, purpose-designed building set in a pleasant rural environment. Secluded green-field sites on the fringes of cities, with good road links and not too far from the homes of consultants and insured patients, have been sought. This has put pressure on Green Belt land which is protected from development, unless the developers can prove a special case which justifies an exception to the general policy. Most of the many applications to develop Green Belt land near the M25 surrounding London for private hospitals have been based on the argument that an institution standing in extensive, well maintained grounds is an appropriate land use since it may preserve or improve the Green Belt. This argument is most likely to succeed when the proposed site is of low visual quality, as in former quarries, waste tips or other derelict land. However, objections from residents based on the strict principle of not allowing new building on Green Belt land have frequently been successful, so ideal sites around the urban fringe are not easily found.

The regional pattern of private hospitals

As with health insurance, the regional distribution of private hospitals is markedly uneven, Mohan has illustrated the disparities using as an indicator the number of beds in private hospitals with operating theatres in 1982. While the average (median) region had 12 private beds per 100,000 population, the four Thames regions and Oxford had the highest rates (from 40 beds for NE Thames down to 19 beds per 100,000 for Oxford). The south-east is the most favourable part of the country for private hospitals as it is for NHS hospitals, but much more so: in 1982, 54 per cent of private beds of this category in England were in the Thames regions. Within the south-east, the West End of London has the greatest concentration, with a focus on Harley Street and the parallel Wimpole Street and Devonshire Place, a small area with an international medical reputation. By contrast, the Northern Region, with 2 beds per 100,000 population, contained only 1 per cent of the private beds. Mohan has drawn attention to these disparities in the context of NHS priorities. While the NHS attempts to redistribute resources from acute to primary and community care and from central London to the more peripheral regions, the private sector is concentrating almost exclusively on secondary acute medical services, especially in London. The forces of the market place are tending to widen rather than diminish existing social and geographical inequalities and the pattern of private health service development is directly counter to that of supposed national priorities.

218

INTERDEPENDENCE OF SOCIAL WELFARE SERVICES

Competitive and complementary health services are not limited to the private sector. There is a wide range of social welfare services whose supply directly affects health and the demand for health services yet which remain outside the control of the NHS. Services provided by social services departments, by district councils and by some voluntary bodies not only complement health services: in some circumstances they may be alternatives. There are inevitably difficulties in co-ordination, and people needing help sometimes fall victim in the gaps between agencies, as well-publicised reports of child deaths due to neglect tragically illustrate.

The social service departments of counties and metropolitan districts are responsible for both residential facilities and field work services which overlap with health services. Residential facilities include old people's homes, day centres and clubs for the elderly, day training centres for the mentally handicapped and some residential and day care services for the physically handicapped and mentally ill. All these facilities relieve NHS hospitals of pressure. Social services day centres can, in fact, be very similar in function to some day hospitals (DHSS, 1981c).

Field work services of social service departments are co-ordinated by social workers, working in teams, each of which covers a geographical patch. Social workers help clients cope with their problems (child care is a particular statutory duty) and arrange for services or benefits to be supplied. They are often instrumental in enabling people to live in the community who might otherwise be in hospital. At least a third of referrals to social workers come from the health service, and other referrals may have a health ingredient. A close working arrangement with primary health teams is desirable and some general practitioners favour the attachment of social workers to the primary care team. Few attachments have actually been made, perhaps because social workers see their responsibilities as being much wider.

Other field services help handicapped people to live at home. Home helps are intended to assist people who cannot manage household work because of illness, infirmities or handicap. In practice they often provide more personal care as well, so there is some overlap with district nurses. Meals on wheels, which delivers a hot meal two or three times a week to a recipient's home, is financed by social services departments but may be actually supplied on an agency basis by the WRVS. Social services departments also provide home alterations and equipment (such as stair lifts, ramps, ground floor lavatories, telephones for the housebound and laundry services for incontinent people), to enable the handicapped and infirm to live at home.

The statutory housing responsibility of district councils gives them, too, an area of common interest with the health service in the provision of specialised housing and wardens to look after groups of residents.

A common thread is that many services have the effect of keeping dependent people out of hospital. In places where social welfare services are less developed, the need for hospital care is correspondingly greater. If one agency is to fill the gaps left by the other, then health services should be strongest where social services are weak and *vice versa*. Whether or not that is the case remains to be resolved. At least, the pattern of social service provision does not appear to follow the 'inverse care law' which has characterised the health services. Bebbington and Davies (1982) have compared the considerable variations in social service provision for the elderly between social services departments in England with an indicator of need in each local authority area. The need indicator was based on estimates of the cost of services considered appropriate for certain target groups of the elderly and was adjusted for local price levels. All the services investigated were found to be positively related to the needs indicator to some extent, but most of the variations in provision remained after different levels of need had been accounted for. Residential provision was the most highly correlated with the need indicator, with a coefficient of 0.62 indicating that 38 per cent of the variations from one local authority to another were related to need and 62 per cent were not. The other types of provision were less related to needs. For meals on wheels the correlation coefficient was 0.47; for home helps 0.33, and for day centre places 0.25 (94 per cent unrelated to needs). Bebbington and Davies concluded that while the current pattern of social service provision shows some responsiveness to need, it does not produce territorial justice.

Joint finance

It is clearly to the advantage of the NHS for social welfare services to be well developed and fully available. Indeed, some policies of the NHS, notably that of discharging mentally ill and mentally handicapped people from hospital, rely on alternative social services provision in the community. Such policies would have little chance of success unless health authorities could offer an incentive for other agencies (which have their own priorities) to co-operate.

The incentive is a pool of NHS money which is available for other agencies to use for schemes which benefit health service clients. It is known as joint finance, and is seen as a catalyst to get health and local authorities to make a real commitment to collaboration and joint planning and encourage other agencies to promote schemes which are expensive solutions for them but cheap in the overall context. Joint finance applied originally to the personal social services, but this excluded schemes such as plans to build a remedial pool for handicapped children in a special school, so it now includes education and housing initiatives. Voluntary organisations and family practitioner committees may also receive joint finance. The procedure is administered locally by a joint consultative committee consisting of representatives of the district health authority and the matching local authority (National Association of Health Authorities, 1983).

Joint finance was originally a pump-priming mechanism to launch local authority capital schemes. The health service money tapered off and finished after five years. For this reason, local authorities were reluctant to agree to projects with substantial continuing costs (for which they would soon become fully responsible) and most schemes concentrated on projects with low running costs. The five-year limit was later extended to 7 years and is now 13 years (DHSS, 1983a), which has opened the way for joint finance to be used for revenue as well as capital support, and makes it a more attractive proposition for recipients. Local authorities, because of their financial difficulties, have little other money available for new developments.

The main beneficiaries of joint finance have been the elderly, allocated about 40 per cent of the resources, and the mentally handicapped, with about 33 per cent (DHSS, 1981a).

Roughly a third of revenue spending has been on residential care, a quarter on day care and two-fifths on domiciliary services and social work. Most of the money, therefore, has not been spent on residential alternatives to hospitals but on services in the community. Whether this has reduced dependence on long-stay hospitals significantly is open to doubt. It has been argued that most of the schemes benefiting from joint finance were planned within local authorities in isolation from the health service, and to meet personal social services priorities rather than NHS priorities (Wistow, 1984).

Locality planning

True co-ordination between the health service and the other agencies whose services overlap can be achieved only between individuals at the local level. An experiment to promote more integrated planning in small geographical areas is currently taking place in the Exeter health district, an area which covers some 300,000 population (Court, 1984). This part of East Devon has been divided into twelve 'localities', which are each catchment areas of the larger centres: Exeter, Okehampton, Crediton, Tiverton, and so on. The localities are an attempt to reflect the natural communities within whose limits most people's activities (work, shopping, school and social activities) take place. They vary in population from 4,000 to 40,000.

The scheme began in 1981 with informal meetings in each locality between health and social service administrators, general practitioners, health workers, social service workers and representatives from housing, pharmacy and voluntary organisations. Staff with similar responsibilities got to know each other and various suggestions for collaboration emerged. One such scheme was the development of multi-purpose day centres to meet the medical and social needs of the elderly in each locality by merging the concepts of the NHS day hospital and the social services day centre. Discussions at the meetings produced detailed proposals which were fed back to the participating statutory bodies. A major scheme to provide geriatric beds centrally in Exeter was altered as a result of the locality groups' initiative and the plan was changed to redistribute the beds in smaller local centres. Support units for the mentally handicapped and homes for confused elderly people are now being provided at locality level.

Since 1984 the locality groups have been established by the health authority, social services department, the local medical committee and other organisations on a more formal basis. The most recent development has been to involve the public in discussions at locality level, with the community health council making the links. According to the associate district administrator, Michael Court, the locality planning approach has changed attitudes in the health authority. It is now much more ready to try to meet the expressed needs of local communities as close to peoples' homes as possible, rather than rationing services from a centralised base. This change of approach might be commended to other health and local government authorities.

PRIORITIES IN THE NHS

Priority services

In England and Wales, the groups acknowledged to be receiving less than their fair share of health services are the elderly, the mentally ill, the mentally handicapped and children. Community and preventive services are also recommended for expansion. The priority groups and services were officially adopted in *The Way Forward* (DHSS, 1977). Since then there have been two more affirmations of priorities, in *Patients First* (DHSS, 1979a) and *Care in Action* (DHSS, 1981d), but these documents were less specific about improvements needed in services and left health authorities to decide on local priorities within a generalised framework of government policy. Scotland has its own priorities (Scottish Health Service Planning Council, 1980) and they are broadly similar. The services in Scotland whose expenditure is intended to grow faster than overall expenditure are those concerned with prevention, services for the multiply deprived, community nursing services, care of the elderly, the elderly with mental disability and services for mental illness, mental handicap and physical handicap. Expenditure on acute hospital services is recommended to remain static or decline in the Scottish priorities document, as it is also for child health services.

Policies and resources

In the event, the priority groups have received few additional resources. The fate of the Court Report, which had been sharply critical of health care provided for children, is typical. The Court Report considered a large quantity of evidence on the health care needs of children, particularly those from deprived social and environmental backgrounds, and presented a long list of recommendations, all designed to secure an integrated child health service accessible to all children and increasingly oriented to prevention (Committee on Child Health Services, 1976). The main organisational change recommended was the establishment of new professional specialists: general practitioner paediatricians, child health visitors and consultant community paediatricians. This suggestion was not accepted by the government of the day, but the Secretary of State did instruct health authorities to strengthen and co-ordinate their child health services along the lines recommended in the report (DHSS, 1978b).

The main thrust was to be a policy of positive discrimination, directing more staff to child health services in the most deprived inner-city areas. Acknowledging that deprived children can be found everywhere, the Secretary of State asserted nevertheless that some localities no bigger than a ward or housing estate contained high concentrations of deprived children, where social and environmental conditions put them at particular risk. Such places might be identified by persistently high perinatal and infant mortality rates and unusually low clinic attendance rates. It was the responsibility of health authorities to concentrate resources on these localities at the expense of more favoured areas. Joint finance was named as a suitable source of funds to strengthen primary care in inner-city areas, by providing extra health visitors, for example, or better accommodation for primary care in health centres. No extra funds were mentioned: authorities were simply urged, in a time of financial constraint, to deploy their resources where they would have most benefit. The objective was 'to raise the health status of the poorer child much nearer to that of the richer child' (DHSS, 1978b). Sceptics might have argued that its achievement was unlikely until more solid resources were committed.

When the Conservative government came into power in

1979 it was pledged to encourage the free market and reduce state expenditure. The Black Report (DHSS, 1980c), which had drawn attention to the widening social inequalities in health and recommended policies to improve the health of lower-income groups and especially children in such groups, was rejected as being too costly in its implications. Expenditure on the National Health Service was restrained by the rigid enforcement of cash limit controls, and the attainment of cash limit targets took precedence over improving or even maintaining services. The small increase in resources which was made available from year to year was consumed by increases in medical costs exceeding the inflation rate and the natural ageing of the population. No additional resources were available to improve the services for the elderly, the mentally ill, mentally handicapped people and children. Improvements in the priority services are therefore dependent on reductions in other services which, as Heller (1978) had predicted, cannot take place without major shifts in the power balance within the NHS.

Under these circumstances there is justification in the criticism that government has paid lip service to the health needs of the least fortunate without taking any effective action (Hubley, 1983). Indeed, some would argue that the action which has been taken (strict control on the NHS combined with permissiveness towards the profit-making sector) has been regressive, in fostering a two-tier system which works against the interests of the most needy and fosters the disparities in services between rich and poor areas (Mohan and Woods, 1985).

Community care

One priority which has been attractive to a government intent on transferring part of the state's responsibility for social welfare back to the individual and the family is the policy of 'community care'. This term has come to describe the practice of moving some people currently occupying hospital beds out of hospital and back into the 'community', whether that is to their home, a relative's home, sheltered housing, social services residential accommodation or even a hotel or boarding house.

The policy is based on the assumption that many people remain in hospital, particularly in long-stay units, not because they need hospital facilities but because there is nowhere else for them to go. These include mentally handicapped people able to look after themselves to a large degree and some mentally ill, physically handicapped or elderly people who could safely leave hospital if special facilities and rehabilitation could be made available. These people and their relatives are said to prefer that they should be looked after 'in the community'. The 'community care' appropriate for many such people will be the residential homes provided by local authorities and voluntary organisations, but not for all. It is possible, for example, that increased support services and day care facilities would enable some mentally handicapped people to move from hospital to ordinary housing or to an adult 'fostering' arrangement. Community health services may also need to be strengthened to make this sort of move viable. From the health service's point of view, subsidising such schemes would be cheaper than maintaining hospital beds and, in the long term, it offers the prospect of closing expensive long-stay hospitals which would not only save running costs but would release the financial resources tied up in buildings and land (DHSS, 1981a). Joint finance has been the vehicle for encouraging social services support and the time limit applying to the health authorities' contribution to joint financial schemes was lengthened specifically to attract more local authority participation (DHSS, 1983a).

Progress in transferring patients out of hospital to some other form of care has been mixed, according to a DHSS study group (DHSS, 1981c). This investigation detected no signs of movement of elderly people out of institutions and back to the community. There have been increases since 1975 in the proportion of elderly people receiving domiciliary or day care services, but these appear to have benefited the less dependent. Those in long-term hospital or residential care have stayed roughly constant in recent years, in spite of DHSS policies that elderly people should be enabled to lead full and independent lives in the community wherever possible. Two trends suggest that the pressure on residential and hospital places has increased. These are the rising age of admission and the increasing dependency of new residents. The researchers conclude that there may have been a tendency to underestimate the core of elderly people who in the long term will require residential care and who might prefer it.

For the mentally ill, government policy is ultimately to phase out the large isolated psychiatric hospitals. This process is under way. The number of long-term mentally ill patients has been falling steadily for about thirty years, with a considerable movement back to the 'community', particularly for the under-65s. Critics have argued that the community services have in many cases been unable to cope and the individuals concerned are too often abandoned with little support. The House of Commons Social Services Committee (1985) heard evidence of barely stable patients who were discharged to bed and breakfast places, seedy lodgings and doss houses. The least fortunate were said to be occupying park benches, gutters and prisons.

For mentally handicapped people there has been a slight decline in the proportion of adults in long-term hospital care and some increases in local authority residential places over the last decade. While the balance is shifting (more slowly than was anticipated ten years ago), the reduction in hospital care is mostly in admissions. Relatively few people have been transferred from hospital to residential care.

The DHSS study group concluded that increases in the provision of community services since 1975 have not been geared directly to providing a genuine alternative for long-standing hospital patients. The Social Services Committee of the House of Commons was in no doubt that patients discharged from hospital were not adequately cared for. In its own words (p. xxii), 'Any fool can close a long-stay hospital: it takes more time and trouble to do it properly and compassionately.' Health care services provided in the community are generally not sufficiently intensive to be an effective substitute for long-term hospital stays, particularly for patients who lack a network of family and friends to support them. Recent increases in community services seem to have mostly benefited people for whom there is no question of long-term institutional care.

The DHSS report was quick to take issue with the commonly held view that community care is more cost effective than residential care. Methodological problems have seriously weakened the attempts that have been made to compare the two. Both costs and effectiveness are difficult to measure and

227

it is not easy to identify comparable groups of people being supported in institutions and at home, since few people over a certain level of dependency actually live on their own in the community. The comparisons that are available do not support the conventional wisdom. Particularly for people living alone, to be cared for at home may be more expensive and perhaps less effective than residential or hospital care. There are cases where, on the face of it, domiciliary care appears to be cheaper, but these may be because the domiciliary support is inadequate.

To decide what kind of care is most appropriate, the individual's physical and mental condition must be considered. The higher the level of dependency, the more appropriate residential care becomes. For the DHSS study team, an equally and possibly more important factor is the availability of informal care from family and friends. Residential types of care are usually called for when this is not available. People who benefit from informal care provided by relatives, on the other hand, are more likely to remain in the community whatever their dependency level. The formal domiciliary services they receive are less expensive than the alternative of residential care, but the balance would be altered if the informal services of their relatives were given a financial value and added to the equation. The financial, social and emotional burdens carried by informal carers can be considerable. The study team recommended that they should be offered more NHS support, for without their contribution the health service would collapse under the strain.

The ageing population

In future years, the health service is faced by an unprecedented demographic challenge. The proportion of people aged 65 and over has been growing during the last thirty years from about 11 per cent of the population of Great Britain in 1951 to about 15 per cent in 1981. While the total number aged 65 and over is expected to dip slightly at the turn of the century before rising again, the number of very old people is predicted to rise throughout the foreseeable future (DHSS, 1984c). The number of people aged 75 and over in England, for example, is expected to reach 3.32 million by 1991. This is a growth of 23 per cent from the 1981 figure.

The rising number of elderly and very elderly people in the population is inevitably swelling the demand for health services, particularly for psychiatric services. A large increase in the incidence of dementia (intellectual failure) and age-related psychiatric disorders is anticipated. About 10 per cent of people of 65 and over are estimated to suffer from dementia and the proportion increases to 20 per cent of those over 80 (NHS Health Advisory Service, 1982). The health care system, which puts most of its resources into the problems of the young and middle-aged, does not appear well equipped to cope. What seems to be needed is an initiative co-ordinated between the NHS and personal social services authorities to give families the right kind of help so that they can look after their confused elderly relatives at home.

The informal care available from families and neighbours is a critical factor. Ominously, there are signs that support for old people at home is in decline. In the years ahead there will be a decrease in the number of women aged 45–60 (who are the main source of support), and increasing employment levels among women suggest that proportionally fewer women will be available to help their elderly parents and relatives cope at home. The growing popularity of retirement migration is separating the generations geographically. Furthermore, expectations about the quality of life and health services are rising, and, without appropriate help, families might find it too difficult to manage and insist that statutory bodies take over. If that were to happen, there is a danger that the National Health Service will be overwhelmed by sheer weight of numbers (NHS Health Advisory Service, 1982).

This particular problem, which threatens to dominate all other priorities in the health service, has a distinctive geographical distribution. The key features are the proportion of elderly in the population, especially the over-75s and over-80s, the numbers who live alone and whether the area is one where family structure is maintained. In the retirement areas, the coastal and rural areas of southern England and north Wales in particular, very large numbers of people will reach advanced age far from the support of their families (Figure 8.2). New towns will also have special problems as the first cohort of residents reaches advanced age at about

Figure 8.2: Population aged 75 and over, 1983: percentages of total population

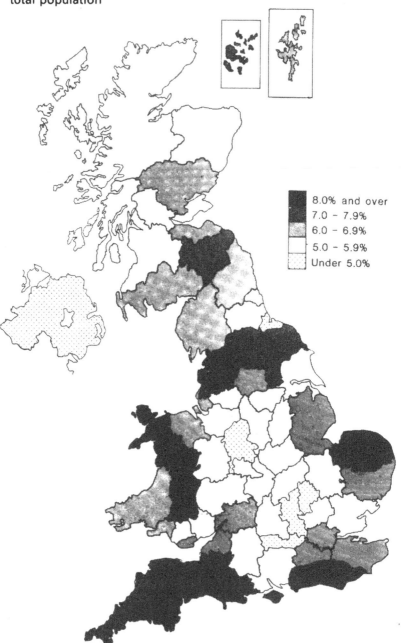

8.0% and over
7.0 - 7.9%
6.0 - 6.9%
5.0 - 5.9%
Under 5.0%

Source: Central Statistical Office, 1985a, p. 48.

the same time. Much of the strain on NHS resources will be concentrated in a few relatively small areas.

RADICAL PROPOSALS FOR CHANGE

Many people working in the National Health Service consider that it is seriously underfunded. Cameron (1985) is a single example of a widespread sense of desperation. They argue that the health needs of the nation are not being met because of the shortage of resources. More money would enable the present services to be provided equitably (many doctors understandably are unhappy about making life or death decisions on how to ration scarce services) and only then could new services, such as those oriented towards prevention, be afforded. The argument is not necessarily for unconstrained spending until all 'treatable' patients can be treated (although perhaps the Hippocratic training and philosophy of many doctors encourage that approach), but for a level of funding which permits the NHS to do its job in a reasonably fair, caring and civilised manner. The proper level can only be determined through the political process, but advocates point out that the United Kingdom spends a lower proportion of its national income on health than any other developed Western nation (with the exception of Greece) and devotes more resources to defence than health.

Prevention versus cure

As was noted in Chapter 2, supplying health services is not the same as promoting good health. In Britain, significant improvements in health have been made since the late eighteenth century, mainly because of rising standards of living and sanitary measures. Better health services have played a comparatively minor part (McKeown, 1979). Improvements in health have been achieved not principally by treating the ill but by reducing the incidence of illness. The same is true today. Medical science directs most of its effort on the health problems which have been presented rather than trying to prevent the problems arising in the first place. While it is very successful in providing symptomatic relief, modern medicine cannot cure the diseases and disabilities which trouble most people from day to day. There are only a few notable exceptions (Acheson and Hagard, 1984). Relieving symptoms might not strike at the root of health problems, but it is undoubtedly of great benefit for individual sufferers. In recent decades there have been considerable developments in scientific methods of treatment. Improvements in surgical techniques and anaesthesia have made operations safer and more likely to be successful, and have led to much shorter stays in hospital than formerly. Advances in pharmacology have supplied physicians with drugs to control the effects of disease. Drugs have also been used successfully to treat the more serious psychiatric disorders. The successes of scientific medicine and surgery hardly need elaboration.

Technological medicine has, however, brought with it a catalogue of unintended illnesses, hospital infections, accidents in the use of dangerous treatments and the adverse effects of drugs (National Association of Health Authorities, 1983). Drugs which attack a particular disease can cause another. The powerful antibiotic chloramphenicol, for example, can induce failure of the white blood cells and eventually the death of the patient. People with cancers which are treated by drugs may run the risk of developing blood cancers and leukaemias which are as likely to cause death as the original malignancy. The increasing prescription of drugs has been accompanied by an increase in addiction to narcotics and barbiturates in pill form.

Surgery, too, has its costs. In Britain, immediate post-operative mortality is estimated at $1-1.5$ per cent, so surgical misadventure may cause up to 20,000 to 30,000 deaths per year. Some operations commonly carried out have been shown to be more dangerous than the consequences of not operating. The surgery of hernias in elderly men is estimated to be four times more risky than not intervening. In addition to the sometimes undesirable physiological effects, possible psychological damage also needs to be considered, particularly in operations like that for breast cancer which apparently produce no measurable increase in life expectancy.

Considerable developments have occurred in chemical and radiological investigation with increasing reliance on expensive high-technology equipment and personnel, but the direct benefits to patients are not always clear. In some fields diagnosis has outstripped treatment; it is possible (at considerable economic cost to the NHS and possibly social cost to the patient) to determine what is wrong but not to remedy it. When a treatment is available, there may be no relationship between the health service input and the eventual outcome. For heart attack patients, for example, differences in the duration of stay in hospital do not seem to have any effect on mortality rates.

Considerations like these have prompted fundamental doubts over the continuing emphasis on expensive interventions once the damage has been done. It is very questionable whether further investment in curative medicine would give good value in terms of reduced premature mortality, serious illness and long-term disablement compared with the alternative of investing in prevention (Acheson and Hagard, 1984). A programme in health education directed at reducing the risks of inadequate diet, smoking and environmental hazards might be expected to produce more lasting results. Instead of devoting most of its energy to short-term hospital care, the NHS might also be more humanely employed helping the people who need support over a long period, whether in hospital or in the community.

Heller (1978) is a critic of the administratively oriented approach which assumes that the way to address the problem of health is to deliver the present package of services more efficiently and fairly. This, in Heller's view, can only contribute marginally to an improvement in health. Various historical and international comparisons have shown that expenditure on health services has little relationship with levels of health. Housing, education, nutrition and income all have stronger relationships with the health of the population than the medical service. The reasons for these relationships are not always well understood, but we should be concerned with changing the aspects of society that can be shown to damage the health of its people and not merely with patching up its casualties. Heller advocates fundamental changes in the present technological illness-oriented health service. He would like to see the curative functions restricted to intervention of proven effectiveness and the caring functions expanded to help the informal systems of looking after people. The medical profession ought increasingly to be teaching the community how to stay healthy and how to cope with simple ailments when they arise, as well as campaigning for a more equitable distribution of the social assets that create good health.

What are the chances of a major reorientation in health service values? Heller would say they are slim, unless the influence of the professional administrators and the medical profession is reduced. Heller sees the National Health Service as a product of the interaction of these two power blocs. The interests of the administrators are in attempting to achieve efficient delivery of services and in controlling the spending of the doctors. The medical profession, on the other hand, resists any attempts to limit its autonomy and protects its orientation towards technological intervention. The tension between them has created a distorted service which has little contact with the needs of the people. Only a shift in power towards the people for whom the service was intended will produce a real improvement in the health service which, to Heller, means a concentration on preventive and domiciliary services rather than on curative and institutional facilities. Even a shift of that magnitude, however, will make little impression on health unless it is part of a much broader egalitarian social policy.

Social inequalities

There is a growing awareness that the single major obstacle to significant improvements in the nation's health is symbolised by low income. Health deprivation is not an isolated phenomenon, but is embedded in a cluster of physical, emotional, social and environmental disadvantages, as has been shown in the longitudinal National Child Development Study described by Wedge (1984). Children who are disadvantaged in terms of health, who are handicapped, absent from school through illness, prone to accidents in the home or not protected as they should be by immunisation and vaccination tend to be those sharing beds in crowded, sub-standard homes, with single parents or unemployed parents living on low incomes, whose progress at school is poor. Association is not the same as causation, of course, and the underlying processes have yet to be exposed. Improvements in health services may help, but they are unlikely to help significantly on their own. Broader measures to improve the living standards of the least fortunate section of the population have a much better chance.

234

The Black Report (DHSS, 1980c) saw the elimination of child poverty as a key point of intervention.

TOWARDS GEOGRAPHICAL EQUITY

The Royal Commission on the National Health Service (1979) declared that the objectives of the NHS are to:

encourage and assist individuals to remain healthy;
provide equality of entitlement to health services;
provide a broad range of services of a high standard;
provide equality of access to these services;
provide a service free at the time of use;
satisfy the reasonable expectations of its users;
remain a national service responsive to local needs.

Two of these objectives – equality of access and responsiveness to local needs – have particular geographical implications. Equality of access in a geographical sense is, of course, impossible to achieve. As long as services are concentrated and the population is dispersed, some people will always have more favourable geographical access than others. The two objectives may also be contradictory, for responsiveness to local needs implies higher levels of access to services where people are most disadvantaged by social or environmental circumstances. Responsiveness to local needs is itself a difficult target at which to aim because of the near impossibility of measuring 'needs' to everyone's satisfaction. Finally, there are conflicts between these and the other objectives. As services develop, high standards are recognised and ultimately achieved not through equality but only when 'centres of excellence' lead the way. Satisfying the expectations of users also tends to work against the aim of equality, both in a social and a geographical sense. Maintaining the present range of services means that there are few resources to devote to assisting people to remain healthy, and so on. It is, of course, not possible to identify a single overriding objective and the NHS must continue to seek the proper balance between several.

Geographical disparities

Efficiency or cost-effectiveness is the criterion used in the National Health Service to distribute resources to particular tasks. Since the Griffiths Report (DHSS, 1984d), general managers have been appointed at all levels with the specific task of promoting efficiency. Using scarce resources in the most cost-effective manner is not the same, however, as providing equality of access for all who need health care. Neither is equality of access (implying that people have the same opportunities) synonymous with equity, or fairness. In an equitable arrangement, people may be treated differently according to their needs. An equal expenditure on health services in two different places can by no means be expected to produce equal health outcomes. An equitable distribution would attempt to match service provision to differences in need by assessing the variations in environmental and social measures known to be related to health. Even then, equal health outcomes are unlikely since health services alone have a relatively small influence on levels of health (Barr and Logan, 1977).

There are three main areas where the geographical distribution of health services does not match the pattern of needs. They are disparities in the provision of services between regions, in remote rural areas and in inner-city districts. The regional disparities have been acknowledged by government and a policy designed to correct them has been in operation for some years. The policy is undoubtedly working, in that differences between regions have diminished, but whether it has significantly increased the equity of access to health care is another matter. Heller (1978) and Eyles *et al.* (1982) have argued that although RAWP creates the appearance of a technical solution to the problem of equity, in reality the policy helps to perpetuate the dominance of the acute hospital sector and fossilises the relative neglect of people with long-term dependency.

In the truly rural parts of Britain a distinctive problem of access to health care is being created by a combination of circumstances. As both general practitioner and hospital services become more centralised, the most rural areas are losing services. Rural people are obliged to travel further afield as their local services are withdrawn. At the same time, the character of truly rural populations is changing. Selective migration is tending to concentrate the old, the poor and the disabled in the most outlying areas. These people are disadvantaged both by their high need for health care and also by their low levels of personal mobility which make it especially difficult to overcome the longer distances to centralised health services. The old, the poor and the disabled appear to receive less health care in very rural areas than they do in more accessible parts of the country.

A geographically dispersed health service is not necessarily preferable to a service that is concentrated in the main centres of population. The trend towards larger units offers more efficient use of scarce resources and an increased quality of service for the majority of consumers. The benefits of centralisation, however, need to be weighed against the costs. This is a task which National Health Service planners are not yet well equipped to do. It will require more information than is available at present on the implications for accessibility of different ways of providing services and practical means for improving access. It will also mean evaluating services from the consumer's point of view and considering broadly defined social costs alongside costs that are internal to the NHS.

The inner-city health care problem is not one of physical access, but it is also peculiarly geographical in that relatively small areas appear to suffer the worst environmental and social conditions. It is tempting, therefore, to advocate a policy of positive discrimination in primary health care for especially deprived areas, as Knox (1979) has done. Once medically deprived areas have been identified using measures of service supply and indices of need, priority treatment might be directed to them. The designated area allowance might be replaced by larger, continuing allowances for fewer doctors in the worst areas. Attempts might be made to persuade inner-city doctors to form partnerships, co-operate more with other welfare workers and use health centres. Whether this could be achieved by investing family practitioner committees with wider powers is debatable. At the heart of the problem are the entrenched professional attitudes of doctors. In the short term, medical education might include more emphasis on the medical and social value of working in deprived areas. Nurse practitioners might be tried as a relatively cheap yet appropriate supplement to general practitioner services in specific

237

places. In the longer term, the control exerted by the medical profession on health care planning should be loosened, so that broader social objectives may be pursued more effectively.

The idea of channelling resources into deprived areas is attractive. Other priority area policies have included the Housing Action Area scheme, the Urban Aid Programme, the Community Development Project, the Comprehensive Community Programme and the Educational Priority Area policy. All such policies run the risk, however, of losing sight of the individuals who are most in need by concentrating on the geographical areas which have the worst average conditions. Averages can be misleading because they may mask considerable variability. In evaluating the effects of education policy it was found that 'the majority of disadvantaged children are not in disadvantaged areas and the majority of children in disadvantaged areas are not disadvantaged' (Barnes and Lucas, 1973). Area policies dilute the effects of aid by helping many individuals who are not underprivileged and missing many more who are. It is not the area which deserves better health services so much as certain people who deserve better health.

Rationing services

The National Health Service is continually subjected to the strain of new demands. Private health services have done little to relieve the most urgent pressure as they have thrived in the places and the branches of medicine with the most resources already. Developments in personal social services offer the possibility of some relief, but genuine co-ordination is still a distant prospect.

A service in which demand always exceeds supply must have effective rationing mechanisms. In Britain, geographical location is a rationing device which helps to determine the allocation of scarce health service resources. When populations are spread over an area and facilities are discrete units with particular locations and inevitable variations in quality, equal access to a uniformly good service is impossible to achieve. If we could be sure that people in the greatest need always receive appropriate levels of care and only those in lesser need are sometimes denied help, then geographical rationing might be accepted as a necessary evil. It seems likely that this is not the case. Studies of the distribution of resources in regions and districts have shown that the 'priority' groups are never the least disadvantaged by geographical variations in health service provision, and they are sometimes the most disadvantaged. Studies of the effects of remoteness and the conditions of the inner-city have shown that the members of the population with the greatest need for health care bear the brunt of comparative disadvantage. The geographical rationing method tends to be regressive: it directs the benefits of health care towards the 'haves' rather than the 'have-nots'. Loosening the geographical constraints on demand, however, cannot be achieved without cost. Either more public money must be devoted to expanding health care or some other form of rationing must be substituted. Attempts to improve access for those in most need can only force the flood gates wider and impose more strain on the service.

References

Abel-Smith, B. (1964) *The Hospitals 1800–1948*, Heinemann, London

—— (1978) *National Health Service: The First Thirty Years*, HMSO, London

Acheson, R. M. and Hagard, S. (1984) *Health, Society and Medicine*, Blackwell, Oxford

Allsop, J. and Lovell, A. (1984) 'Why Some People Just Never Make it to the Surgery', *Health and Social Service Journal, 94*, 463–4

Andrew, H. R. and Harris, M. R. (1978) 'The Traffic Attraction of West Midlands Hospitals', *Traffic Engineering and Control*, July, 331–5

Ashford, J. R. (1978) 'Regional Variations in Dental Care in England and Wales', *British Dental Journal, 145*, 275–83

Automobile Association Technical Services (1977) *Schedule of Estimated Running Costs: April 1977*, Automobile Association, Basingstoke

Avery-Jones, F. (1976) 'The London Hospital Scene', *British Medical Journal, 2*, 1046–9

Bain, S. (1983) 'Health Care Problems in North-West Sutherland: A Peripheral Region of Scotland' in N. D. McGlashan and J. R. Blunden (eds.), *Geographical Aspects of Health*, Academic Press, London, pp. 223–40

Barnes, J. A., Merchant, J. A. and Webster, N. (1974) *The Ambulance Service: Performance, Standards and Measurement*, Statistics and Operational Research Unit, Cranfield Institute of Technology, Cranfield

Barnes, J. H. and Lucas, H. (1973) 'Positive Discrimination in Education: Individuals, Groups and Institutions' in J. Barnes (ed.), *Educational Priority*, Department of Education and Science, London

Barr, A. and Logan, R. F. L. (1977) 'Policy Alternatives for Resource Allocation', *Lancet, 1*, 994–6

Bebbington, A. C. and Davies, B. (1982) 'Patterns of Social Service Provision for the Elderly' in A. M. Warnes (ed.), *Geographical Perspectives on the Elderly*, John Wiley, Chichester

Beer, T. C., Goldenberg, E., Smith, D. S. and Mason, A. S. (1974) 'Can I Have an Ambulance, Doctor?' *British Medical Journal, 1*, 226–8

Bentham, C. G. (1984) 'Mortality Rates in the More Rural Areas of England and Wales', *Area, 16*, 219–26

—— (1986) 'Proximity to Hospital and Mortality from Motor Vehicle Traffic Accidents', *Social Science and Medicine*, forthcoming

—— and Haynes, R. M. (1985) 'Health, Personal Mobility and the Use of Health Services in Rural Norfolk', *Journal of Rural Studies*, *1*, 231–9

Bevan, A. (1945) *Memorandum by the Minister of Health to the Cabinet, 5 October 1945*, Public Record Office, CAB 129/3

Birrell, D. and Williamson, A. (1983) 'Northern Ireland's Integrated Health and Personal Social Service Structure' in A. Williamson and G. Room (eds.), *Health and Welfare States of Britain*, Heinemann, London

Black, D. (1983) 'Need, Demand, Supply', *Effective Health Care* (Amsterdam), *1*, 3–5

—— (1984) *Investigation of the Possible Increased Incidence of Cancer in West Cumbria: Report of the Independent Advisory Group, Chairman: Sir Douglas Black*, HMSO, London

Bloor, M., Horobin, G., Taylor, R. and Williams, R. (1978) *Island Health Care: Access to Primary Services in the Western Isles*, Institute of Medical Sociology, University of Aberdeen, Occasional Paper No. 3

Bolden, K. J. (1981) *Inner Cities*, Royal College of General Practitioners, Occasional Paper No. 19

Bowen, P. G. (1976) 'Patient Transport – Can the Demand be Met?' *Health and Social Services Journal*, *3*, 18

Bradley, J. E., Kirby, A. M. and Taylor, P. J. (1976) *Distance Decay and Dental Decay*, University of Newcastle-upon-Tyne, Department of Geography, Seminar Paper No. 31

Bradshaw, J. S. (1972) 'A Taxonomy of Social Need' in G. McLachlan (ed.), *Problems and Progress in Medical Care. Essays on Current Research, Seventh Series*, Oxford University Press, London

Brennan, M. E. and Clare, P. H. (1980) 'The Relationship between Mortality and Two Indicators of Morbidity', *Journal of Epidemiology and Community Health*, *34*, 134–8

British Medical Journal (editorial) (1977) 'Making Better Use of Ambulances', *British Medical Journal*, *2*, 1242–3

Brodsky, H. and Hakkert, A. S. (1983) 'Highway Fatal Accidents and Accessibility of Emergency Medical Services', *Social Science and Medicine*, *17*, 731–40

Brown, D. B. (1979) 'Countermeasures Evaluation: A Study of Emergency Medical Services', *Journal of Safety Research*, *11*, 37–41

Brown, R. G. S. (1973) *The Changing National Health Service*, Routledge and Kegan Paul, London

Butler, J. R., Bevan, J. M. and Taylor, R. C. (1973) *Family Doctors and Public Policy*, Routledge and Kegan Paul, London

—— and Knight, R. (1975a) 'The Choice of Practice Location', *Journal of the Royal College of General Practitioners*, *25*, 496–504

——, —— (1975b) 'Designated Areas: A Review of Problems and Policies', *British Medical Journal*, *2*, 571–3

——, —— (1976) 'Medical Practice Areas in England: Some Facts and Figures', *Health Trends*, *8*, 8–12

Button, J. H. (1984) 'Wales Today', *Hospital and Health Services Review*, 110–15

Buxton, M. J. and Klein, R. E. (1975) 'Distribution of Hospital Provision: Policy Themes and Resource Variations', *British Medical Journal, 1*, 345–9

——, —— (1978) *Allocating Health Resources: A Commentary on the Report of the Resource Allocation Working Party*, Research Paper No. 3, Royal Commission on the NHS, HMSO, London

Calder, G. (1983) 'Organisation of Pharmacy Practice in Scotland', *IPM Journal, 5*, 12–13

Cameron, S. (1985) Letter to the Editor, *Guardian*, 13 March

Caple, L. M. (1976) 'Hospital Car Service', *Health and Social Services Journal*, 14 February, 308

Carmichael, C. L. (1983) 'General Dental Service Care in the Northern Region', *British Dental Journal, 154*, 337–9

Cartwright, A. (1964) *Human Relations and Hospital Care*, Routledge and Kegan Paul, London

—— and Anderson, R. (1981) *General Practice Revisited*, Tavistock Publications, London

Cassel, J. (1977) 'The Relation of the Urban Environment to Health: Towards a Conceptual Frame and a Research Strategy' in L. E. Hinkle and W. C. Loring (eds.), *The Effect of the Man-Made Environment on Health and Behaviour*, US Dept. of Health, Education and Welfare, Atlanta, Georgia, 129–42

Cavenagh, A. J. M. (1978) 'Contribution of General Practitioner Hospitals in England and Wales', *British Medical Journal, 2*, 34–6

Central Health Services Council (1969) *The Functions of the District General Hospital*, HMSO, London

Central Statistical Office (1985a) *Regional Trends, 20*, HMSO, London

—— (1985b) *Social Trends, 15*, HMSO, London

Chilvers, C. (1978) 'Regional Mortality 1969–73', *Population Trends, 11*, 16–20

Clayton, S. (1984) *Long Distances to Hospital*, Dept. of Social Administration, Lancaster University

Coates, B. E. and Rawstron, E. M. (1971) *Regional Variations in Britain: Selected Essays in Economic and Social Geography*, Batsford, London

Cobb, J. S. and Miles, D. P. B. (1983) 'Estimating List Inflation in a Practice Register', *British Medical Journal, 287*, 1434–6

Cochrane, A. L. (1976) 'The London Hospitals Scene', *British Medical Journal, 2*, 1384

——, St Leger, A. and Moore, F. (1978) 'Health Service Inputs and Mortality Output in Developed Countries', *Journal of Epidemiology and Community Health, 32*, 200–5

College of Health (1985) *Guide to Hospital Waiting Lists*, College of Health, London

Collings, J. S. (1950) 'General Practice in England Today', *Lancet*, 25 March, 555–85

Collins, E. and Klein, R. (1980) 'Equity and the NHS: Self-reported

Morbidity, Access and Primary Care', *British Medical Journal*, *281*, 1111–15

Committee on Child Health Services (1976) *Fit for the Future: Report of the Committee on Child Health Services* (Court Report), Cmnd 6684, HMSO, London

Cook, P. J. and Walker, R. O. (1967) 'The Geographical Distribution of Dental Care in the United Kingdom', *British Dental Journal*, *122*, 441–7, 494–9, 551–8

Cooper, M. H. (1975) *Rationing Health Care*, Croom Helm, London

Court, M. (1984) 'Grassrooting for Better Health Services', *NAHA News*, November, 6

Craft, A. W., Openshaw, S. and Birch, J. (1984) 'Apparent Clusters of Childhood Lymphoid Malignancy in Northern England', *Lancet*, *2*, 96–7

Craig, J. (1983) 'The Growth of the Elderly Population', *Population Trends*, *32*, 28–34

Cross, K. W. and Turner, R. D. (1974) 'Factors Affecting the Visiting Pattern of Geriatric Patients in a Rural Area', *British Journal of Preventive and Social Medicine*, *28*, 133–9

Cullis, J. G., Foster, D. P. and Frost, C. E. B. (1981) 'Met and Unmet Demand for Hospital Beds', *Revue d'Epidemiologie et Sante Publique*, *29*, 155–66

Culyer, A. J. (1976) *Need and the National Health Service*, Martin Robertson, London

Currie, E. (1984) 'Plans and Patients: Crossed Lines?' *Health and Social Service Journal*, *94*, 1496–7

Curtis, S. E. (1984) *Intra-urban Variations in Health and Health Care: The Comparative Need for Health Care Survey of Tower Hamlets and Redbridge*, vol. 1, Adult Morbidity and Service Use, Dept. of Geography and Earth Science, Queen Mary College, University of London

Dennis, N. (1984) 'Voting with their Feet', *Health and Social Service Journal*, *94*, 465

Dental Strategy Review Group (1981) *Towards Better Dental Health: Guidelines for the Future*, DHSS, London

Department of the Environment (1981) *Circular 2/81*, DoE, London

Department of Health and Social Security (1962) *Report of a Sub-committee on Accident and Emergency Services* (Platt Report), HMSO, London

—— (1970a) *Annual Report 1969*, HMSO, London

—— (1970b) *Domiciliary Midwifery and Maternity Bed Needs*: Report of a sub-committee of the Standing Medical and Midwifery Advisory Committee of the Central Health Services Council (Peel Report), HMSO, London

—— (1970c) *The Organisation and Staffing of Operating Departments*: Report of a joint sub-committee of the Standing Medical Advisory Committee and the Standing Nursing Advisory Committee of the Central Health Services Council (Lewin Report), HMSO, London

—— (1971a) *Better Services for the Mentally Handicapped*, Cmnd

4683, HMSO, London

—— (1971b) *The Organisation of Group Practice*: Report of a sub-committee of the Standing Medical Advisory Committee, HMSO, London

—— (1974) *Community Hospitals: Their Role and Development in the National Health Service*, HSC (IS) 75, DHSS, London

—— (1976) *Sharing Resources for Health in England*: Report of the Resource Allocation Working Party, HMSO, London

—— (1977) *Priorities in the Health and Social Services: The Way Forward*, HMSO, London

—— (1978a) *Ambulance Services: Operational Control and Use*, HC (78) 45, DHSS, London

—— (1978b) *Court Report on Child Health Services*, HC (78) 5, DHSS, London

—— (1979a) *Patients First*, HMSO, London

—— (1979b) *Primary Health Care: Health Centres and Other Premises*, HC (79) 8, DHSS, London

—— (1979c) *Transport Act 1978: Public Transport Planning, Community Bus Services and Social Car Schemes*, HC (79) 5, DHSS, London

—— (1980a) *Health Centre Policy*, HC (80) 6, DHSS, London

—— (1980b) *Report of the Advisory Group on Resource Allocation*, DHSS, London

—— (1980c) *Report of the Working Group on Inequalities in Health* (Black Report), DHSS, London

—— (1980d) *Structure and Management*, HC (80) 8, DHSS, London

—— (1981a) *Care in the Community: A Consultative Document on Moving Resources for Care in England*, HC (81) 9, DHSS, London

—— (1981b) *Contractual Arrangements with Independent Hospitals and Nursing Homes*, HC (81) 1, DHSS, London

—— (1981c) *Report of a Study on Community Care*, DHSS, London

—— (1981d) *Care in Action: A Handbook of Policies and Priorities for the Health and Personal Social Services in England*, HMSO, London

—— (1982) *Health and Personal Social Services Statistics 1982*, HMSO, London

—— (1983a) *Care in the Community and Joint Finance*, HC (83) 6, DHSS, London

—— (1983b) *Community Health Service Dental Staff: England and Wales National Tables*, Statistics and Research Division, DHSS, London

—— (1983c) *Independent Sector Hospitals, Nursing Homes and Clinics in England*, DHSS, London

—— (1983d) *Pharmaceutical Services: Implementation of the Report of the National Joint Committee of the Medical and Pharmaceutical Professions on the Dispensing of NHS Prescriptions in Rural Areas*, HN (FP) (83) 9, DHSS, London

—— (1983e) *Remuneration of General Medical Practitioners*, HN (FP) (83) 19, DHSS, London

—— (1983f) *Supra-regional Services*, HN (83) 36, DHSS, London

—— (1984a) *Fares to Hospital: Leaflet H11*, HMSO, London
—— (1984b) *General Practitioner Deputising Services*, HC (FP) (84) 2, DHSS, London
—— (1984c) *The Health Service in England: Annual Report 1984*, HMSO, London
—— (1984d) *Implementation of the NHS Management Inquiry Report*, HC (84) 13, DHSS, London
—— (1984e) *Remuneration of Essential Small Pharmacies*, HN (FP) (84) 28, DHSS, London
—— (1984f) *Temporary Scheme for Incentives to Group Practice in Inner City Areas: Statement of Fees and Allowances*, HN (FP) (84) 14, DHSS, London
—— (1985a) *Health and Personal Social Service Statistics for England, 1985 Edition*, HMSO, London
—— (1985b) *The Health Service in England: Annual Report 1985*, HMSO, London
DHSS Working Party on Under-Doctored Areas (1979) *Draft Report*, DHSS, London
Department of Health and Social Services, Northern Ireland (1978) *Proposals for the Allocation of Revenue Resources for Health and Personal Social Services*: Report of the Working Group on Revenue Resource Allocations to Health and Social Services Boards in Northern Ireland, Department of Health and Social Services, Belfast
—— (1981) *The Structure and Management of Health and Personal Social Services in Northern Ireland*, HSS (P) 1/81, Department of Health and Social Services, Belfast
Department of Transport (1983) *National Travel Survey 1978/9 Report*, HMSO, London
Dunnell, K. and Cartwright, A. (1972) *Medicine Takers, Prescribers and Hoarders*, Routledge and Kegan Paul, London
—— and Dobbs, J. (1982) *Nurses Working in the Community*, HMSO, London
Eyles, J., Smith, D. M. and Woods, K. J. (1982) 'Spatial Resource Allocation and State Practice: The Case of Health Service Planning in London', *Regional Studies*, 16, 239–53
—— and Woods, K. J. (1983) *The Social Geography of Medicine and Health*, Croom Helm, London
Exeter and District Community Health Council (1983) *Medical Services in Rural Areas*, Exeter and District Community Health Council, Exeter
Farmer, R. D. T. (1984) 'Patients Like to be an Emergency', *Health and Social Services Journal*, 94, 466
Fearn, R. M. G. (1983) *The Role of the Branch Surgery in Accessibility to Primary Health Care in Rural Norfolk*, unpublished PhD thesis, University of East Anglia, Norwich
—— , Haynes, R. M. and Bentham, C. G. (1984) 'Role of Branch Surgeries in a Rural Area', *Journal of the Royal College of General Practitioners*, 34, 488–91
Ferrar, H. P. (1977) 'Mortality, Morbidity and Resource Allocation',

Lancet, 1, 1054

——— , Moore, A. and Stevens, G. C. (1977) 'The Use of Mortality Data in the Report of the Resource Allocation Working Party (HMSO 1976)', *Public Health, 91*, 289–95

Forest, D. and Sims, P. (1982) 'Health Advisory Services and the Immigrant', *Health Trends, 14*, 10–13

Forster, D. P. (1977) 'Mortality, Morbidity and Resource Allocation', *Lancet, 1*, 997–8

Forsythe, J. M. (1983) 'Health Care in English Cities', *World Hospitals, 19*, 38–41

Fox, J. (1977) 'Occupational Mortality 1970–72', *Population Trends, 9*, 8–15

Freeman, H. L. (1978) 'Mental Health and the Environment', *British Journal of Psychiatry, 132*, 113–24

Gardner, M. J., Winter, P. D. and Barker, D. J. P. (1984) *Atlas of Mortality from Selected Diseases in England and Wales 1968–1978*, John Wiley, London

Gatrell, A. (1985) 'Vaccination Take-up in Salford: An Ecological Study', paper presented to the IBG and AAG Joint Symposium in Medical Geography, Nottingham

Geary, K. (1977) 'Technical Deficiencies in RAWP', *British Medical Journal, 1*, 1367

Gibbons, J. (1978) 'The Mentally Ill' in P. Brearley, J. Gibbons, A. Miles, E. Topliss and G. Woods (eds.), *The Social Context of Health Care*, Blackwell, London

Grant, J. A. (1984) 'Contribution of General Practitioner Hospitals in Scotland', *British Medical Journal, 288*, 1366–8

Gray, A. McI. (1982) 'Inequalities in Health. The Black Report: A Summary and Comment', *International Journal of Health Services, 12*, 349–80

Grime, L. P. and Whitelegg, J. (1982) 'The Geography of Health Care Planning: Some Problems of Correspondence between Local and National Policies', *Community Medicine, 4*, 201–8

Gruer, R. (1972) *Outpatient Services in the Scottish Border Counties*, Scottish Home and Health Department, Edinburgh

Harrison, A. and Gretton, J. (1984) *Health Care UK 1984: An Economic, Social and Policy Audit*, Policy Journals, Hermitage

Hart, J. T. (1971) 'The Inverse Care Law', *Lancet, 1*, 405–12

Haynes, R. M. (1985) 'Regional Anomalies in Hospital Bed Use in England and Wales', *Regional Studies, 19*, 19–27

——— and Bentham, C. G. (1979a) 'Accessibility and the Use of Hospitals in Rural Areas', *Area, 11*, 186–91

———, ——— (1979b) *Community Hospitals and Rural Accessibility*, Saxon House, Farnborough

———, ——— (1979c) 'Measuring the Accessibility of Health Services', *Hospital and Health Services Review, 75*, 118–21

———, ——— (1982) 'The Effects of Accessibility on General Practitioner Consultations, Out-patient Attendances and In-patient Admissions in Norfolk, England', *Social Science and Medicine, 16*, 561–9

247

Heller, T. (1978) *Restructuring the Health Service*, Croom Helm, London

Hencke, D. (1985) 'Ambulancemen Fail 999 Standards', *Guardian*, 10 July, 4

Hillman, M. and Whalley, A. (1975) 'Land Use and Travel', *Built Environment Quarterly, 1*, 105–11

Hobbs, M. S. T. and Acheson, E. D. (1966) 'Perinatal Mortality and the Organization of Obstetric Services in the Oxford Area in 1962', *British Medical Journal, 1*, 499–505

Hopkins, E. G., Pye, A. M., Solomon, M. and Solomon, S. (1968) 'The Relation of Patients' Age, Sex and Distance from Surgery to the Demand on the Family Doctor', *Journal of the Royal College of General Practitioners, 16*, 368–78

House of Commons (1974) *Expenditure Committee, Fourth Report, Session 1973–74: Accident and Emergency Services*, HMSO, London

—— (1975) *Accident and Emergency Services: Government Observations on the Fourth Report of the House of Commons Expenditure Committee*, Cmnd 5886, HMSO, London

—— (1984) *Parliamentary Debates* (Hansard), Sixth Series, 59, 2 May 1984, HMSO, London

House of Commons Social Services Committee (1985) *Community Care with Special Reference to Adult Mentally Ill and Mentally Handicapped People, Second Report*, HMSO, London

Howat, J. G. M. and Kontny, E. L. (1977) 'What Price the Ambulance? A Survey of Psychiatric Day-patient Transport', *British Medical Journal, 2*, 1298–9

Hubley, J. (1983) 'Poverty and Health in Scotland' in G. Brown and R. Cook (eds.), *Scotland: The Real Divide*, Mainstream Publishing, Edinburgh

Hunt, S. and McEwen, J. (1980) 'The Development of a Subjective Health Indicator', *Sociology of Health and Illness, 2*, 231–46

——, McKenna, S., McEwan, J. and Papp, E. (1981) 'The Nottingham Health Profile: Subjective Health Status and Medical Consultations', *Social Science and Medicine, 15A*, 221–9

Hyde, A. (1984) 'Lincoln Gets Carried Away', *Health and Social Services Journal, 94*, 558–9

Institute of Health Service Administrators (1984) *The Hospitals and Health Services Yearbook 1984*, Institute of Health Service Administrators, London

Jarman, B. (1981) *A Survey of Primary Care in London*, Occasional Paper 16, Royal College of General Practitioners, London

—— (1983) 'Identification of Under-privileged Areas', *British Medical Journal, 286*, 1705–8

—— (1984) 'Underprivileged Areas: Validation and Distribution of Scores', *British Medical Journal, 289*, 1587–92

—— (1985) Personal correspondence

Joseph, A. E. and Phillips, D. R. (1984) *Accessibility and Utilization: Geographical Perspectives on Health Care Delivery*, Harper and Row, London

Knox, P. L. (1978) 'The Intraurban Ecology of Primary Medical Care: Patterns of Accessibility and their Policy Implications', *Environment and Planning A*, *10*, 415–35

—— (1979a) 'The Accessibility of Primary Care to Urban Patients: A Geographical Analysis', *Journal of the Royal College of General Practitioners*, *29*, 160–8

—— (1979b) 'Medical Deprivation, Area Deprivation and Public Policy', *Social Science and Medicine*, *13D*, 111–21

—— and Pacione, M. (1980) 'Locational Behaviour, Place Preferences and the Inverse Care Law in the Distribution of Primary Medical Care', *Geoforum*, *11*, 43–55

Leavey, R. (1983) *Inequalities in Urban Primary Care: Use and Acceptability of GP Services*, DHSS Research Unit Working Paper, Department of General Practice, University of Manchester

Levitt, R. and Wall, A. (1984) *The Reorganised National Health Service*, 3rd edn, Croom Helm, London

London Health Planning Consortium (1979) *Acute Hospital Services in London*, HMSO, London

London Health Planning Consortium Study Group (1981) *Primary Health Care in Inner London*, London Health Planning Consortium, London

Lumb, R. (1983) *Access to Medical Services in Rural Northumberland*, Community Council of Northumberland, Morpeth

Mair, R. (1983) 'Health Services' in J. English and F. Martin (eds.), *Social Services in Scotland*, Scottish Academic Press, Edinburgh

Maynard, A. and Ludbrook, A, (1980) 'Applying Resource Allocation Formulae to Constituent Parts of the U.K.', *Lancet*, *1*, 85–8

——, —— (1983) 'The Allocation of Health Care Resources in the United Kingdom' in A. Williamson and G. Room (eds.), *Health and Welfare States of Britain*, Heinemann, London

McCarthy, M. (1983) 'New Directions for Primary Health Care in the Inner City', *Hospital and Health Services Review*, *79*, 11–13

McGirr, E. M. (1984) 'Planning Health Care in Scotland', *British Medical Journal*, *289*, 776–8

McKeown, T. (1979) *The Role of Medicine*, Blackwell, Oxford

——, Cross, K. W. and Keating, D. M. (1971) 'Influence of Hospital Siting on Patient Visiting', *Lancet*, *2*, 1082–6

Medical Practices Committee (1985a) *Annual Report 1984*, Medical Practices Committee, London

—— (1985b) Personal communication

Medical Services Review Committee (1962) *A Review of the Medical Services in Great Britain* (Chairman, Sir A. Porritt), Social Assay, London

Ministry of Health (1962) *A Hospital Plan for England and Wales*, Cmnd 1604, HMSO, London

Mohan, J. (1984a) 'Planners, Politicians and the Development of the Hospital Services of Newcastle upon Tyne 1948–1969', *Environment and Planning C*, *2*, 471–84

—— (1984b) *Spatial Aspects and Planning Implications of Private Hospital Developments in South East England, 1976–84*.

Occasional Paper, Dept. of Geography, Birkbeck College, University of London

—— and Woods, K. J. (1985) 'Restructuring Health Care. The Social Geography of Public and Private Health Care under the British Conservative Government', *International Journal of Health Services*, *15*, 197–215

Morrell, D. C., Gage, H. G. and Robinson, N. A. (1970) 'Patterns of Demand in General Practice', *Journal of the Royal College of General Practitioners*, *19*, 331–42

Moseley, M. J. (1979) *Accessibility: The Rural Challenge*, Methuen, London

—— and Packman, J. (1983) *Mobile Services in Rural Areas*, School of Environmental Sciences, University of East Anglia, Norwich

Moyes, A. (1977) 'Accessibility to General Practitioner Services on Anglesey: Some Trip Making Implications', paper presented to the Institute of British Geographers Conference, Newcastle upon Tyne

National Association of Health Authorities (1983) *NHS Handbook*, National Association of Health Authorities, London

National Federation of Women's Institutes (1977) *Health Services in Rural Areas. A Survey of the Views of a Selected Sample of Women's Institute Members*, National Federation of Women's Institutes, London

NHS Health Advisory Service (1982) *The Rising Tide: Developing Services for Mental Illness in Old Age*, NHS Health Advisory Service, Sutton, Surrey

Norman, A. (1977) *Transport and the Elderly: Problems and Possible Action*, National Corporation for the Care of Old People, London

Noyce, J., Snaith, A. H. and Trickey, A. J. (1974) 'Regional Variations in the Allocation of Financial Resources to the Community Health Services', *Lancet*, *1*, 554–8

Office of Population Censuses and Surveys (1983a) *Census 1981 National Report: Great Britain*, HMSO, London

—— (1983b) *General Household Survey 1981*, HMSO, London

O'Mullane, D. M. and Robinson, M. E. (1977) 'The Distribution of Dentists and the Uptake of Dental Treatment by Schoolchildren in England', *Community Dentistry and Oral Epidemiology*, *5*, 156–9

Open University (1977) *Inequality within Nations: Health and Inequality* (D302 Unit 13), Open University, Milton Keynes

—— (1985) *Caring for Health: Dilemmas and Prospects* (Health and Disease, U205, Book VIII), Open University Press, Milton Keynes

Palmer, S., West, P., Patrick, D. and Glynn, M. (1979) 'Mortality Indices in Resource Allocation', *Community Medicine*, *1*, 275–81

Parkin, D. (1979) 'Distance as an Influence on Demand in General Practice', *Epidemiology and Community Health*, *33*, 96–9

Patient Transport Services Working Party (1981) *Patient Transport Services*, Trent Regional Health Authority, Sheffield

Pharmaceutical Society of Great Britain (1985) Personal correspondence

250

Phillips, D. R. (1979a) 'Spatial Variations in Attendance at General Practitioner Services', *Social Science and Medicine*, *13D*, 169–81

—— (1979b) 'Public Attitudes to General Practitioner Services: A Reflection of an Inverse Care Law in Intra-urban Primary Medical Care?' *Environment and Planning A*, *11*, 815–24

—— (1980) 'Spatial Patterns of Surgery Attendance: Some Implications for the Planning of Primary Health Care', *Journal of the Royal College of General Practitioners*, *30*, 688–95

—— (1981) *Contemporary Issues in the Geography of Health Care*, Geo Books, Norwich

—— and Radford, J. P. (1985) 'Any Fool Can Close a Long-Stay Hospital: Deinstitutionalization and Community Care for the Mentally Handicapped – A Study of South West England', paper presented to the Joint IBG and AAG Medical Geography Conference, University of Nottingham

Pocock, S. J., Shaper, A. G., Cook, D. G., Packham, R. F., Lacey, R. F., Powell, P. and Russell, P. F. (1980) 'British Regional Heart Study: Geographic Variations in Cardiovascular Mortality, and the Role of Water Quality', *British Medical Journal*, *280*, 1243–9

Political and Economic Planning (1944) *Medical Care for Citizens*, Planning Broadsheet 222, Political and Economic Planning, London

Population Statistics Division OPCS (1982) 'The New District Health Authorities', *Population Trends*, *29*, 13–14

Radical Statistics Health Group (1977) *RAW(P) Deals*, Radical Statistics Health Group, London

Redfern, P. (1982) 'Profile of our Cities', *Population Trends*, *30*, 21–32

Richardson, J. and Stroud, P. (1975) *Passenger Requirements in an Automated Transport System*, Supplementary Report 175 UC, Transport and Road Research Laboratory, Crowthorne

Rickard, J. H. (1976a) *Cost-Effectiveness Analysis of the Oxford Community Hospital Programme*, Health Services Evaluation Group, Dept. of the Regius Professor of Medicine, University of Oxford

—— (1976b) 'Per Capita Expenditure of the English Area Health Authorities', *British Medical Journal*, *1*, 299–300

Ritchie, J., Jacoby, A. and Bone, M. (1981) *Access to Primary Health Care*, HMSO, London

Robinson, C., Lennon, M. A. and O'Mullane, D. M. (1980) 'The Influence of Family Ties on the Practice Location of Dentists in Three Contrasting Groups of Towns', *Journal of Dental Research*, *59*, Special Issue D, 1836

Royal Commission on the National Health Service (1979a) *Report*, Cmnd 7615, HMSO, London

—— (1979b) *Access to Primary Care*, Research Paper No. 6, HMSO, London

Sadler, J. and Whitworth, T. (1975) *Reserves of Nurses*, HMSO, London

Scarrott, D. M. (1978) 'Changes in the Regional Distribution of

General Dental Service Manpower', *British Dental Journal, 144,* 359–63

Scottish Health Service Planning Council (1980) *Priorities for Health Care in Scotland,* HMSO, Edinburgh

Scottish Home and Health Department (1967) *General Medical Services in the Highlands and Islands* (Birsay Report), Cmnd 3257, HMSO, Edinburgh

—— (1977) *Scottish Health Authorities Revenue Equalisation: Report of the Working Party on Revenue Resource Allocation,* HMSO, Edinburgh

Senn, S. J. and Shaw, H. (1978) 'Resource Allocation: Some Problems in Applying the National Formula to Area and District Revenue Allocations', *Journal of Epidemiology and Community Health, 32,* 22–7

Steering Group on Health Services Information, Working Group G (1983) *Patient Transport Services,* DHSS, London

Taylor, P. J. and Carmichael, C. L. (1980) 'Dental Health and the Application of Geographical Methodology', *Community Dentistry and Oral Epidemiology, 8,* 117–22

Thexton, A. J. and McGarrick, J. D. (1983) 'The Geographical Distribution of Recently Qualified Dental Graduates (1975–1980) in England, Scotland and Wales', *British Dental Journal, 154,* 71–6

Todd, J. E. (1975) *Children's Dental Health in England and Wales 1973,* HMSO, London

Townsend, P. (1981) 'Toward Equality in Health through Social Policy', *International Journal of Health Services, 11,* 63–75

—— and Davidson, N. (1982) *Inequalities in Health: The Black Report,* Penguin Books, Harmondsworth

Tuckett, D. (ed.) (1976) *An Introduction to Medical Sociology,* Tavistock Publications, London

Wedge, P. (1984) 'Combating Deprivation and Enhancing Childhood', *Health and Hygiene, 5,* 47–51

Welsh Consumer Council (1979) *Getting Primary Care on the NHS,* Welsh Consumer Council, Cardiff

Welsh Office (1977) *Report of the Steering Committee on Resource Allocation in Wales,* HMSO, Cardiff

—— (1983) *Health and Personal Social Services Statistics for Wales 1982,* Welsh Office, Cardiff

West, R. R. and Lowe, C. R. (1976) 'Regional Variations in Need for and Provision and Use of Child Health Services in England and Wales', *British Medical Journal, 2,* 843–6

Whewell, J., Marsh, G. N. and McNay, R. A. (1983) 'Changing Patterns of Home Visiting in the North of England', *British Medical Journal, 286,* 1259–61

While, A. (1982) 'Primary Health Care in the Inner City: Time for a New Approach?' *Health Visitor, 55,* 116–17

Whitelegg, J. (1982) *Inequalities in Health Care: Problems of Access and Provision,* Straw Barnes, Retford, Nottinghamshire

Williams, A. (1974) 'Need as a Demand Concept (with Special Reference to Health)' in A. J. Culyer (ed.), *Economic Policies and Social Goals*, Martin Robertson, London

Wistow, G. (1984) 'Joint Finance and Community Care' in A. Harrison and J. Gretton (eds.), *Health Care UK 1984: An Economic, Social and Policy Audit*, Policy Journals, London

Wood, J. (1983a) 'Are the Problems of Primary Care in Inner Cities Fact or Fiction?' *British Medical Journal*, *286*, 1109–12

—— (1983b) 'Are General Practitioners in Inner Manchester Worse Off than Those in Adjacent Areas?' *British Medical Journal*, *286*, 1249–52

Woods, K. J. (1982) 'Social Deprivation and Resource Allocation in the Thames Regional Health Authorities' in Health Research Group (ed.), *Contemporary Perspectives on Health and Health Care*, Occasional Paper No. 20, Dept. of Geography, Queen Mary College, University of London

Yudkin, J. S. (1978) 'Changing Patterns of Resource Allocation in a London Health District', *British Medical Journal*, *2*, 1212–15

Index